TROUBLED
PARTNERSHIP

An Assessment of U.S.-Japan Collaboration on the FS-X Fighter

Mark Lorell

Project AIR FORCE

Prepared for the
United States Air Force

RAND

PREFACE

This report presents a synthesis and overview of the key findings and recommendations of a RAND research study on U.S.-Japan cooperative development of the FS-X fighter. It focuses on issues of technology transfer and the long-term implications of cooperative development programs for the American aerospace industry. It is meant to be read in conjunction with a companion volume (Lorell, 1995), which provides a detailed case study and history of the FS-X program from its origins through 1993.

This report emerged from a RAND research project conducted in the early 1990s on collaboration with Asian allies on military aircraft research and development (R&D). The Resource Management and System Acquisition Program of RAND's Project AIR FORCE initiated this research, which was sponsored by the United States Air Force.

This report and its companion volume are intended to assist U.S. government officials in formulating better policies and strategies for effective military technology collaboration with Japan and other allies. They should also be of interest to the general reader who is concerned with U.S. industrial competitiveness and maintaining America's preeminence in defense R&D.

The views expressed in this report are those of the author and do not reflect the official policy or position of the Department of Defense or the U.S. government.

PROJECT AIR FORCE

Project AIR FORCE, a division of RAND, is the Air Force federally funded research and development center (FFRDC) for studies and analyses. It provides the Air Force with independent analyses of policy alternatives affecting the development, employment, combat readiness, and support of current and future aerospace forces. Research is being performed in three programs: Strategy, Doctrine, and Force Structure; Force Modernization and Employment; and Resource Management and System Acquisition.

Project AIR FORCE is operated under Contract F49620-91-C-0003 between the Air Force and RAND.

CONTENTS

FIGURES

ACKNOWLEDGMENTS

From 1992 thorough 1994, the author conducted extensive interviews with scores of current and former U.S. government officials directly involved in the FS-X program. These included representatives from various U.S. Air Force offices, Air Force and Navy R&D laboratories, Air Force liaison offices in Japanese Industry and the Japan Defense Agency, the Office of the Secretary of Defense, the Defense Security Assistance Agency, the Defense Technology Security Administration, the Department of Commerce, the Department of State, the U.S. Embassy in Tokyo, the Mutual Defense Assistance Office, the Government Accounting Office, and the Office of Technology Assessment. On the U.S. industry side, numerous managers and engineers at Lockheed Fort Worth and Lockheed Nagoya (formerly General Dynamics), General Electric engines, and Westinghouse were interviewed. Related interviews were conducted with representatives of Boeing and McDonnell-Douglas.

To gain a better understanding of Japanese government policies, the author interviewed many officials at the Japan Defense Agency Policy and Equipment Bureaus, the Air Self-Defense Force Air Staff Office, the Technical Research and Development Institute, the Ministry of International Trade and Industry, the Ministry of Foreign Affairs, and the Defense Research Center. Industry views were sought through extensive discussions with managers and engineers at Mitsubishi Heavy Industries, Kawasaki Heavy Industries, Fuji Heavy Industries, and Mitsubishi Electronics Corporation.

Many individuals in the government, the armed forces, and the aerospace industries of both the United States and Japan provided critical assistance in the preparation of this book. They are far too numerous to thank individually here. However, several individuals particularly stand out for the extensive help and support they provided in the course of this research effort. These include Maj Craig Mallory (USAF); Elizabeth Marini, Defense Technology Security Administration, CAPT Tim Horner (USN, Ret), Defense Security Assistance Agency; and Charline Gormly, Office of the Secretary of the Air Force. I am deeply grateful to them, and to all the other individuals on both sides of the Pacific who shared so generously of their time and knowledge.

Numerous RAND colleagues also made invaluable contributions to this effort. Foremost among them were George Donohue, former RAND Vice President and Director of Project AIR FORCE, and David Lyon, former Vice President for External Affairs, who encouraged this research from its very inception. Francis Fukuyama and Robert Lempert formally reviewed this manuscript and provided numerous substantive comments and criticism of great merit. The author also appreciates the patience, support, and intellectual contributions of Project AIR FORCE managers, particularly Michael Kennedy, Dennis Smallwood, Jeff Drezner, and Robert Roll. Cindy Kumagawa, Book Program Director, offered important guidance on preparing this work for publication. A special debt of gratitude is owed to Laura Zakaras, Communications Analyst. Without her constant support and encouragement, as well as substantial intellectual input, this effort could not have been completed.

ABBREVIATIONS

APA	Active phased array
ASDF	Japanese Air Self-Defense Force
CCV	Control configured vehicle
CFRP	Carbon fiber-reinforced plastic
CRS	Congressional Research Service
dem/val	Demonstration and validation
DoD	Department of Defense
EFA	European Fighter Aircraft
EW	Electronic warfare
FBIS-EAS	Foreign Broadcast Information Service-East Asia Daily Report
FBW	Fly-by-wire
FCC	Flight control computer
FS-X	Fighter Support-Experimental
GaAs	Gallium arsenide
GAO	General Accounting Office
GD	General Dynamics
HiMAT	Highly maneuverable technology
JDA	Japanese Defense Agency
JPRS-JST	Joint Publications Research Service-Japan Science & Technology
KHI	Kawasaki Heavy Industries

LFWC	Lockheed Fort Worth Company
MBB	Messerschmitt-Bölkow-Blohm
MELCO	Mitsubishi Electronics Corporation
MHI	Mitsubishi Heavy Industries
MITI	Ministry of International Trade and Industry
MMIC	Monolithic microwave integrated circuit
MoA	Memorandum of Agreement
MoU	Memorandum of Understanding
R&D	Research and development
TKF	*Taktisches Kampflugzeug* [German tactical combat aircraft]
TMD	Theater missile defense
T/R	Transmit and receive
TRDI	Technical Research and Development Institute

Chapter One

PROGRAM HISTORY: PROTRACTED DISPUTES AND MIXED OUTCOMES

The Fighter Support Experimental (FS-X) program is the largest cooperative international military aircraft development program in U.S. history. It is also the largest aerospace weapon system research and development (R&D) project undertaken by the Japanese government since the Second World War. It calls for the cooperative U.S.-Japan development of a new fighter based on the Lockheed (formerly General Dynamics) F-16C Block 40 fighter aircraft for the Japanese Air Self-Defense Force (ASDF).[1] The FS-X will replace the existing Mitsubishi F-1 support fighter. Its primary role will be antiship attack, with a secondary air superiority role.

Mitsubishi Heavy Industries (MHI) is the prime contractor for the FS-X program, in association with Kawasaki Heavy Industries (KHI) and Fuji Heavy Industries on the Japanese side, and Lockheed Fort Worth Company (LFWC) on the American side. The official R&D program budget now stands at about $3.3 billion,[2] although the actual R&D program cost is probably higher. The Japanese government is paying for the entire program. Japanese industry receives 60 percent of the official R&D budget, while the remainder goes to U.S. industry, with LFWC receiving about three-quarters of the U.S. share and most of the rest going to General Electric for the engines. A production run of 130 aircraft is currently planned.

---

[1]In early 1993, Lockheed purchased the General Dynamics (GD) Fort Worth fighter division, the original developer of the F-16. This division was renamed the Lockheed Fort Worth Company (LFWC).

[2]¥330 billion at ¥100 to the dollar.

Although in many ways a precedent-setting example of U.S.-Japan cooperation in defense technology, the FS-X program has also proven to be a long and difficult experiment in international collaboration for both sides. Beginning with the Japanese decision to launch an indigenous fighter development program in 1985, the United States and Japan engaged in nearly five years of difficult and often contentious negotiations before agreeing upon basic terms so that R&D could begin. Throughout these negotiations, the United States pressed for several objectives: to prevent the Japanese from developing an all-new fighter aircraft on their own, to promote Japanese procurement of an existing or minimally modified U.S. fighter, to limit Japanese access to U.S. technological know-how, and to ensure U.S. access to Japanese defense-related technology. Japan, on the other hand, has been motivated chiefly by the desire to ensure development of its own national fighter or, if forced to collaborate, to maximize modification of an existing fighter design and ensure Japanese control over design and development. At this point, in 1995, the year of the prototype's first test flight and ten years after the start of negotiations, the program outcomes are becoming clear: While broadly based on the F-16, the FS-X constitutes a virtually all-new world-class fighter aircraft developed largely by the Japanese. And U.S. industry, which won access to Japanese-developed technology after bitter and prolonged disputes, has shown little interest in that technology.

How and why did the FS-X program evolve away from the original Pentagon conception of a minimally modified American fighter toward something that approximates the indigenous national fighter that the Japanese sought all along? Why did the United States fight so hard to limit Japanese access to its technology, if the result would be to encourage Japan to engage in more indigenous development? And why did the United States insist on gaining access to Japanese technology in which U.S. industry had little genuine interest? This report tries to answer those questions. Its companion volume (Lorell, 1995) offers a detailed case history of the program and provides extensive documentation supporting the conclusions and observations included in this document.

OVERVIEW: A DECADE OF DIFFICULT NEGOTIATIONS

Collaboration with Japan was initially proposed by the Pentagon in 1986 to head off the Japanese initiative to develop a national fighter. Elements within Japanese industry and the security establishment had been interested in developing greater autonomy in Japan's military R&D capabilities and arms production as far back as the 1950s. By 1985, these elements had convinced the Japanese government to support development of a world-class indigenous fighter, code-named FS-X, which could take its place as the modern Zero of the post-war era. Later that year, the Pentagon moved into action to block this move. Initially, the Pentagon proposed licensed production of the F-16, but it soon shifted strategies and suggested collaborative development of the FS-X based on a "minimally modified" General Dynamics F-16C or McDonnell-Douglas F-18. It took three years of difficult negotiations before the two sides decided upon the final terms of a compromise agreement. They finally completed and signed a Memorandum of Understanding (MoU) in November 1988, which laid out the basic structure of the joint program. In January 1989, the principal contractors also signed an industry agreement.

The terms of the deal granted Japanese industry full R&D leadership, although the program is monitored at the government level by a joint U.S.-Japan Technical Steering Committee chaired by two coequal general officers, one representing the U.S. Air Force, the other the ASDF.

U.S. industry agreed to provide Japan with an extensive technical data package for the F-16C, which included more than 32,000 technical drawings. The U.S. side also agreed to transfer a considerable amount of developmental data to explain the design approach. However, all these data were carefully examined by U.S. government experts, and "sanitized" by the removal of any militarily or commercially sensitive information. U.S. industry also agreed to provide around 70 engineers, who would be located at MHI's facilities in Nagoya, Japan, to help explain the technical data package, take part in the development of the FS-X, and ensure flowback of Japanese technology to the United States.

In return, Japan paid $60 million to LFWC for the F-16 technical data package and agreed to pay the American company a license fee of about $500,000 for every aircraft manufactured. The Japanese government also guaranteed that U.S. industry would receive 40 percent of the R&D budget and a similar percentage of the production budget.

Perhaps the most controversial and difficult elements of the deal involved flowback and access to Japanese technology. The government agreements stipulated that Japan would provide all technology applied to the FS-X that was "essentially derived" from U.S. technology to the United States free of charge. Latter agreements required that this free "flowback" of "derived" technology would take place in an expeditious manner.[3] Perhaps most important, the original government agreements designated and defined as "derived" *all* technology applied to the FS-X, with the exception of only four Japanese avionics systems. The agreements included procedures, however, for the Japanese to petition for removing a specific technology or subsystem from the derived category if indigenous development could be proven.[4]

The four Japanese indigenously developed avionics systems originally designated as nonderived technology are the Mitsubishi Electronic Corporation's (MELCO's) active phased-array (APA) fire-control radar, Japan Aviation Electronics' inertial reference system, the MELCO-developed mission computer, and MELCO's integrated electronic warfare system. The FS-X agreements guarantee that the Japanese government will not block U.S. access to these "nonderived" systems and their embedded technologies. However, U.S. companies that seek access are required to negotiate with the Japanese contractor who owns the technology to determine the cost and terms of transfer. The U.S. government primarily plays the role of facilitator.

Negotiations, however, did not end with the agreements signed in early 1989. Soon after the signing, a bitter public debate over the merits of the deal exploded in Congress and the press. As Figure 1 shows, a number of follow-on agreements had to be negotiated over the next several years as a result of this controversy and other problems. Opponents of the agreement in early 1989 called

---

[3]In official FS-X program terminology, *flowback* refers to "derived" technology, and *access* refers to "nonderived" technology.

[4]The industry-level agreements contain a more expansive definition of FS-X technology flowback requirements.

the FS-X program a "technological giveaway" to one of our most powerful economic competitors, claiming that the U.S. government was transferring commercially valuable aerospace technology through military aerospace collaboration programs. They raised concerns over the trade deficit, the need to maintain the international dominance of the U.S. aircraft industry, and the potential loss of American jobs. Ultimately, the domestic debate—which was largely between the Pentagon and the Department of State on the one hand and Congress and the Department of Commerce on the other—forced the Bush administration to insist on clarifications to the agreement, causing considerable anger and frustration in Japan.

Even after the Japanese accepted certain clarifications to the original agreements in the spring of 1989, Congress continued to argue over the program for many months. Flowback and access to Japanese technology became much more important politically as symbols of greater reciprocity in the flow of technology between the United States and Japan. The Pentagon position in support of the program stressed the value of U.S. access to certain Japanese technology, particularly the manufacturing processes for the

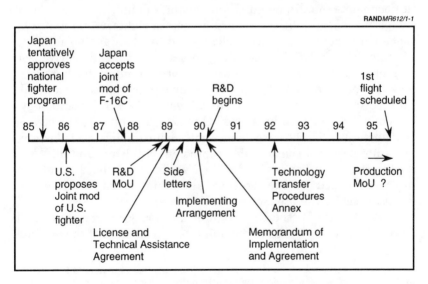

Figure 1—The FS-X Program: Ten Years of Contentious Negotiations

Mitsubishi-developed wing box designed as a single-piece, cocured, carbon-fiber composite (CFC) structure and for the MELCO gallium arsenide (GaAs) monolithic microwave integrated circuit (MMIC) chips for the transmit-receive (T/R) modules mounted on the antenna array of the APA radar.

Congress finally approved FS-X collaboration by a margin of just one vote in September 1989.[5] However, the congressional battle increased the political sensitivity of the transfer of U.S. technology and U.S. access to Japanese technology, and disputes continued to simmer. Many additional months of negotiations were necessary to further define and clarify the "clarifications" negotiated by the Bush administration. The Memorandum of Implementation and Agreement was negotiated in 1990. Following that, more prolonged negotiations took place over a Technology Transfer Procedures Annex that was to clarify U.S. access rights to FS-X technology further and to help establish mechanisms for transferring it to the United States. That document was finally signed in early 1992. However, problems related to access to Japanese technology continued to plague the program.

The various FS-X agreements relating to technology transfer that emerged from these protracted negotiations are highly legalistic documents containing numerous complex definitions, technology categories, and conditions regarding access rights. As a result of these agreements, each technology used by the Japanese on the FS-X had to be defined according to several criteria: derived or nonderived, military or nonmilitary, foreground or background, and Japanese Defense Agency (JDA) owned or contractor owned. This resulted in *twelve* separate categories of FS-X technology, each of which differently affected U.S. access rights and the means of transfer. It is not surprising that such complicated rules and procedures caused considerable confusion, problems, and disputes during the early phases of R&D. Disputes over these questions have continued to the present and delayed the start-up of negotiations for the production MoU.

Actual R&D for the FS-X fighter did not get under way until April 1990, nearly a year and a half after the signing of the original

---

[5]President Bush had vetoed a revision of the original agreements formulated by Congress. The Senate needed a two-thirds majority to override the veto, but came up one vote short.

agreements and almost five years after the beginning of the original negotiations. In the years that followed, U.S. policymakers focused on the issue of protecting U.S. technology from the Japanese and guaranteeing access and flowback of Japanese technology. Meanwhile, extensive changes to the baseline F-16 design were being quietly carried out in Japan.

FROM F-16 TO RISING SUN FIGHTER

Despite years of haggling and stacks of signed agreements, the FS-X program is not meeting many of the initial expectations of the Pentagon negotiators when the program was agreed to in 1987. The single most important shortfall of the program is that it has evolved away from the original Pentagon concept of a minimally modified F-16 to a virtually all-new Japanese-developed fighter broadly based on the F-16.

Why is this important? FS-X is providing Japanese industry with an entrée into the highly exclusive world club of developers of advanced fighter aircraft weapon systems, one of the most potent conventional weapons in existence. This has potentially major long-term implications for the U.S. military aerospace industry and for U.S. security policy.

In outward appearance, the FS-X still closely resembles the F-16, as shown in Figure 2. But appearances can be deceptive. Over 95 percent of F-16 engineering drawings are being changed for the FS-X.[6] Mitsubishi has essentially used the existing F-16 design as a reference guide and starting point for its own extensive design excursions that go far beyond the routine engineering changes normally associated with typical modification programs.

The FS-X wing is an all-new Japanese design that is 25 percent larger in area than the F-16 wing. Its structure and materials are based on a Japanese-developed cocured CFC process. The horizontal stabilizer is also a newly designed composite structure, about 20 percent larger than the F-16 tail plane. Japanese-developed stealth technology is being applied to the airframe. The center

---

[6]In early 1995, Dr. Vernon Lee, Lockheed Vice President for Japan, told a reporter that "Well over 50% of its [FS-X's] parts are new, but from a design concept point of view, less than 50% of the aircraft is new." This does not necessarily contradict the claim that about 95 percent of the F-16 drawings have been changed. See Mecham (1995).

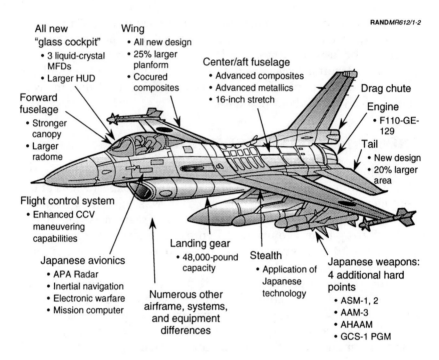

RAND*MR612/1-2*

All new
"glass cockpit"
• 3 liquid-crystal
  MFDs
• Larger HUD

**Forward
fuselage**
• Stronger
  canopy
• Larger
  radome

Wing
• All new design
• 25% larger
  planform
• Cocured
  composites

Center/aft fuselage
• Advanced composites
• Advanced metallics
• 16-inch stretch

Drag chute

Engine
• F110-GE-
  129

Tail
• New design
• 20% larger
  area

Flight control system
• Enhanced CCV
  maneuvering
  capabilities

Japanese avionics
• APA Radar
• Inertial navigation
• Electronic warfare
• Mission computer

Landing gear
• 48,000-pound
  capacity

**Numerous other
airframe, systems,
and equipment
differences**

Stealth
• Application of
  Japanese
  technology

**Japanese weapons:
4 additional hard
points**
• ASM-1, 2
• AAM-3
• AHAAM
• GCS-1 PGM

Figure 2—Main Differences Between the FS-X and the F-16

fuselage is 10 inches longer and has new structures and materials.
The nose and canopy are changed, as is the landing gear. There is
a Japanese-developed "glass" cockpit with three liquid-crystal flat-
panel displays and numerous other new items. In addition to the
four primary indigenous Japanese avionics systems, at least 40 or
more important subsystems and major components are Japanese
developed. Virtually all of the FS-X avionics will be Japanese-
developed component systems or modified versions of F-16 systems.
The FS-X will be armed with many indigenous munitions, including air-to-air and antiship missiles. And the list goes on.

Because Japanese industry has dominated the entire design,
development, and integration process, it is gaining extensive experience in the crucial R&D and system integration process for developing modern fighters. Although the experience gained will not
fully equal that of a true *ab initio* indigenous fighter development,
it will be considerable. Japanese companies are not likely to learn

much about building the next generation of commercial airliners from FS-X, but they are likely to be well positioned to compete on the next generation of fighter aircraft and subsystems.[7]

The ultimate irony is that the FS-X collaboration agreement forced on Japan in 1987 probably ended up in many ways representing a better deal from the perspective of Japanese industry than a purely indigenous development would have. By using F-16 drawings and technical data instead of a totally new and untested design started from scratch, Japanese engineers were able to adopt a more incremental, lower-risk approach of experimenting with interesting variations on a proven baseline design. The program thus provides Japanese industry a lower-risk and less costly approach to honing its design and integration skills, while providing plenty of flexibility for the further development and application of its indigenous technologies and subsystems. As one knowledgeable U.S. industry expert intimately involved with the R&D effort summed it up:

> [F]rom the Japanese perspective, the FS-X development program has been cost-effective for them. They essentially developed an indigenous fighter for approximately $3 billion—quite a deal, all things considered.[8]

Moreover, while U.S. industry is receiving roughly 40 percent of the overall R&D budget, it has little direct involvement in the developmental process for the most interesting and important new technologies, such as the APA radar. LFWC is designing and developing the aft fuselage, the leading-edge flaps, the stores management system, two software test stations, and avionics test equipment. Lockheed is manufacturing several of the left wing boxes for the prototype test program. Even with respect to the co-cured composite wing box, however, Lockheed is in a "leader-follower" role in which MHI does the development primarily on its

---

[7]Some observers have rightly noted that basing the FS-X on the F-16 reduces Japanese experience in the "front end" aspect of system engineering compared to a fully indigenous development program, in which engineers carry out the transition from operational requirements to performance requirements for individual configuration items.

[8]Letter to the author, 1994. Other recent foreign fighter R&D programs, such as EF2000 and Rafale, are running at $10–12 billion or more.

own, and the end results are transferred to the American company. This transfers the "know-how," but not necessarily all the "know-why." Ironically, Lockheed has the design and development lead on the aft fuselage, the only major fuselage structure constructed primarily out of conventional metals as opposed to advanced composites.[9]

LESSONS LEARNED

This report sets out what went wrong with the FS-X program and what is at stake in the future evolution of the program. It discusses the problem of imposing collaboration on a reluctant ally, the lack of clear leadership in the United States that led to conflicting policy goals, and misguided assumptions behind U.S. policy on technology transfer and access. It also recommends steps that can be taken to improve the process of collaboration and the outcome of the program. The history of difficulties of negotiating the FS-X agreement has strained relations between the two countries, and the same difficulties now threaten to spill over into negotiations for a production agreement. In fact, the United States and Japan could still be on a collision course that could result in the disruption or cancellation of the program. The United States needs to take steps now to prevent such an outcome. The most important economic, technological, and political benefits of the program for the United States depend on the FS-X entering into series production.

The FS-X experience illustrates that cooperative military development programs carry the potential for significantly aiding a foreign country that is trying to increase its independent military R&D capabilities. In the long run, if such programs are not carefully managed, they can lead to a reduction of U.S. influence over the security policies of important allies and can help establish competitive foreign defense industries that may undermine the U.S. defense industrial base.

---

[9]Early in the R&D program, GD tried to convince the Japanese to support development of the aft fuselage using advanced composite materials. This effort was blocked by both the Japanese—who argued such an effort would cost too much—and American government officials—who expressed concern over the transfer of advanced U.S. composite technology to Japanese industry.

Chapter Two

WHAT WENT WRONG?

Our research identified five general problem areas and policy errors that caused the FS-X R&D program to produce only mixed results and led to many of the disputes and difficulties that have plagued the joint effort. First, the American side imposed cooperative development on a reluctant partner. Second, the FS-X program should have been structured to provide greater U.S. influence over the final design configuration and technological evolution of the aircraft. Third, the U.S. government underestimated Japan's military R&D capabilities. Fourth, the U.S. government did not formulate and implement a single coordinated strategy toward collaboration with Japan that harmonized both U.S. military and economic objectives. Fifth, U.S. policy on technology transfer and access was fundamentally flawed. Successful collaboration in the future will depend on an understanding of the mistakes of the past. As one senior Japanese defense expert wryly noted: "It [the FS-X program] will be a success if we never repeat it again!"[1]

COLLABORATION IMPOSED

Japan: Reluctant Partner

Many Japanese government and industry officials did not want to collaborate with the United States on the FS-X and resented being forced to do so. As in the case of many other sovereign nations, much of the Japanese security establishment has historically sup-

---

[1]Statement by Tetsuo Tamama, Japan Defense Research Council, at the U.S./Japan Economic Agenda's *Conference on High Technology Policy-Making in Japan and the United States: Case Studies of the HDTV and FSX Controversies*, Georgetown University, Washington, D.C., June 8, 1993.

ported development of a first-line national fighter on an indigenous basis. For decades following the end of the Second World War, the Japanese defense industry and elements within the JDA and the services pressed hard to gain government approval for the indigenous development of a world-class "Rising Sun" fighter.[2] Beginning in the early 1960s, the Japanese industry and the military R&D establishment progressively built up the national technological capabilities to develop an all-Japanese indigenous fighter through dedicated military R&D programs, learning through licensed production of advanced American fighters and by "spinning on" advanced technologies from the commercial sector.

The proponents of a national fighter had often been stymied by widespread concerns in the Japanese political establishment over the high costs of indigenous development and the likelihood of an unfavorable reaction from the United States. But in mid-1985, after decades of frustration, the advocates of indigenous development appeared to be on the verge of achieving their long-sought goal. In June of that year, the Japanese government gave a tentative go-ahead for development of an indigenous fighter, the FS-X.

America's leading fighter contractors at that time—GD and McDonnell-Douglas—strongly opposed this turn of events and immediately began lobbying the U.S. government to stop a national fighter program in Japan. Japan had long been a highly lucrative market for American aerospace defense contractors. Every major fighter procured by the ASDF since the Second World War—as well as most other military aircraft acquired by the ASDF—had been license-produced or was a directly purchased version of American-designed and -developed aircraft.[3] The American companies feared the loss of a key market and worried about the emergence over the long run of a highly competitive Japanese military aerospace industry.

---

[2]In the early 1970s, Japanese industry developed the Mitsubishi F-1 support fighter. This fighter, however, was a modification of an existing trainer aircraft, the Mitsubishi T-2. It closely resembles the Anglo-French SEPECAT Jaguar attack aircraft and uses the same engines as the European aircraft. Most foreign observers dismissed the F-1 as a relatively low-performance aircraft that lacked the combat capability of other contemporary first-line fighters.

[3]With the exception of the Mitsubishi F-1 and several other support and transport aircraft.

The U.S. departments of Defense and State were also very concerned, but primarily for strategic and military-political reasons. They believed Japan would waste its limited defense funds on developing a fighter that would prove to be far more expensive and less capable than an existing U.S. fighter bought "off the shelf" or modestly modified. Furthermore, they worried that a Japanese-developed fighter would not be fully interoperable with U.S. fighters. On a longer-term strategic level, U.S. officials opposed indigenous development because they believed it would lead to a more fully capable and autonomous Japanese defense industrial base, which in turn could bolster a more independent Japanese security policy less amenable to U.S. influence. Concerns were also expressed about the effects on regional stability if Japan's neighbors, such as Korea and China, became uneasy over the emergence of a more capable and independent Japanese defense industry.

As a result of these concerns, U.S. officials brought increasing political pressure to bear on the Japanese government over a two-year period beginning in late 1985. Although Pentagon officials originally hoped to separate the FS-X issue from general trade disputes between the two countries, this proved to be impossible. However, as American pressure intensified during this period, Japanese resistance stiffened. American officials quickly concluded that Japan could never be convinced to license-produce or purchase directly an off-the-shelf U.S. fighter. Consequently, the Department of Defense (DoD) proposed a compromise: cooperative modification of an existing U.S. fighter. Japan resisted this solution for many months, but U.S. pressure ultimately became overpowering. In November 1987, Japan agreed to the cooperative development of the FS-X based on a GD F-16C fighter.

Although Japan had finally agreed to a cooperative modification program in late 1987, it still remained a reluctant partner. This was because key players in the Japanese security establishment remained bitterly opposed to collaboration based on a lightly modified U.S. fighter. The highest political levels of the Japanese government had imposed collaboration for political reasons on the supporters of indigenous development, who were concentrated in industry, the Equipment Bureau of the Defense Agency, and the Technical Research and Development Institute (TRDI), which conducts all military R&D in Japan. They had worked long and hard to build up a more autonomous defense industry capability and deeply resented the political imposition of collaboration. They had expended too much time and effort laying the technological and political foundations for a national fighter development effort to let their dream die because of a political decision made by Japan's leadership.

Japanese Counterstrategy: Maximize Indigenous Development

Once the basic agreement had been accepted, the Japanese political leadership left the negotiation of the details of collaboration in the hands of the supporters of indigenous development. They in turn developed and implemented a clever counterstrategy to the American proposal aimed at maximizing modifications to the baseline F-16 and the application of Japanese technology developments and subsystems, while minimizing U.S. control over the technical evolution of the R&D effort.

Indeed, by late 1987, Japanese officials had already made considerable progress in implementing this strategy. Although Pentagon officials believed they had agreed to a rather modest joint R&D program for a "lightly modified" F-16, GD had already accepted Japanese demands for the inclusion of a considerable number of Japanese design changes, technology, and subsystems. Nonetheless, the resulting U.S. modification proposal that formed the basis for the FS-X, called the SX-3, was still thought to be generally similar to the GD-developed Agile Falcon F-16 modification proposal for the U.S. Air Force. Furthermore, it was widely believed that U.S. industry would dominate the development process and that the U.S. government would exercise considerable influence over the technological evolution of the program. This view held sway because American industry had far more expertise in the complex process of fighter R&D and because American officials intended to establish a joint government oversight body.

Nonetheless, the Japanese supporters of indigenous development proved remarkably successful in carrying out their strategy of transforming the "lightly modified" F-16 into something approaching the Rising Sun fighter. They took advantage of the vagueness of the original FS-X agreements between the United States and Japan, which precisely defined neither the final design and configuration of the FS-X nor the details of joint program structure and implementation. U.S. and Japanese industry had agreed on the general outlines of the cooperative modification proposal—including an enlarged wing, a stretched fuselage, maneuvering canards, and Japanese avionics—but this remained basically a broad design concept that required considerable further refinement. This vagueness provided the Japanese with the maneuvering room necessary to transform the FS-X design and its technological content.

Two broad U.S. policy errors provided the Japanese with the basic opportunity to carry out this strategy. The first involved the failure to press hard enough for licensed production of a U.S. fighter. The second was the failure to control the technological and design evolution of the FS-X once the Japanese accepted cooperative development based on the F-16.

THE LACK OF U.S. INFLUENCE OVER THE TECHNOLOGICAL EVOLUTION OF THE FS-X

While its ultimate objective of stopping indigenous development was justifiable from the American perspective, DoD's strategy for implementation was seriously flawed. Once cooperative R&D commenced, the U.S. side was largely powerless to prevent Japanese industry from transforming a "minimally modified" F-16 into the Rising Sun fighter. This was primarily because JDA was vested with final configuration control authority, since the Japanese government was paying the entire bill for R&D.

The United States could have structured the program in one of two different ways to meet the Pentagon's original objectives more effectively:

- It is plausible to argue that the U.S. side could have pushed much harder on the political level for licensed production of a U.S. aircraft by Japan. However, this strategy entailed some risk that Japan might reject the U.S. position and move ahead with indigenous development.

- The U.S. side could have structured a more genuinely collaborative joint R&D program that included significant U.S. government funding and specific U.S. design and technology objectives meant to contribute to U.S. weapon systems. This could have provided far more U.S. influence over the technological evolution of the FS-X.

The following two subsections discuss these options in more detail.

Failure to Press Hard for Licensed Production

Senior Reagan administration officials were correct in 1989 when they testified to Congress that Japan could not be expected under any circumstances to purchase a U.S. fighter off the shelf to fulfill the FS-X requirement. Japanese industry had an understandable requirement for a new production program in the 1990s to keep its workers and extensive facilities employed after F-15 licensed production came to an end. By the 1980s, no leading industrialized country anywhere in the world was willing to purchase U.S. fighters off the shelf. Even many newly industrializing countries, such as Turkey and South Korea, routinely demanded technology transfer and significant industrial participation in major foreign weapon purchases.

However, the claims made by Reagan administration officials that Japan would have never agreed to license-produce a U.S.-developed fighter are somewhat less convincing. While the Japanese military R&D establishment clearly expressed its strong opposition to licensed production, there is no compelling evidence that the United States government forcefully and persistently pressed this option on the senior Japanese political leadership. There is no question that the Japanese military R&D establishment bitterly opposed licensed production—but hardly more than they opposed cooperative development based on a U.S. fighter. The senior Japanese political leadership, however, appears to have been far more vulnerable to U.S. political pressure. If the Japanese political leadership had been induced to accept licensed production, there is little that the working-level officials in the military R&D establishment could have done to subvert U.S. objectives during the actual implementation of the program. Licensed production would have meant joint U.S.-Japanese manufacture of an existing U.S. design, making it virtually impossible for significant design and equipment changes to take place on the Japanese side without full U.S. cooperation.

From the very beginning of the government-level discussions on FS-X in the middle of 1985, Secretary of Defense Caspar Weinberger rejected the option of directly pressuring the Japanese, on the grounds that such tactics could disrupt the overall U.S.-Japan security relationship. In the Cold War environment of the mid-1980s, senior Pentagon officials adamantly opposed linking trade and security issues. Senior DoD officials legitimately concluded that U.S. strategic and military interests were best served by preserving the strong and close security relationship with Japan. In their view, it made little sense to threaten or undermine that relationship over the question of Japan's next fighter. As a result, DoD decided to adopt a strategy of "pressuring without pressure." The U.S. side would provide Japan with cost and performance data on U.S. aircraft in the hopes that such data would encourage the Japanese to reject indigenous development on their own accord because of the high costs of the latter approach.

Yet this strategy was doomed to failure, because it was based on an incorrect understanding of Japanese motives. The Japanese military R&D establishment was perfectly willing to pay considerably more for a less capable fighter, if development of that fighter would contribute significantly to expanding the skill base and military R&D capabilities of the Japanese defense industry.

By early 1986, few Pentagon officials believed the Japanese would ever select licensed production if left to their own devices. U.S. contractors had offered the Japanese various proposals for licensed production, but these elicited little interest. JDA and the Air Staff Office flatly rejected this option and made their views well known to the American side.

Nevertheless, Weinberger's policy of "pressuring the Japanese by not pressuring them" was faithfully adhered to at least until the spring of 1987. There are few indications that high-level administration officials ever applied significant political pressure of any sort on the Japanese during this period to license-produce an existing U.S. fighter, other than arguing that indigenous development would be prohibitively expensive. To the contrary, as early as January 1986, the U.S. side suggested the option of cooperative development of a modified American fighter. Thus, the ultimate form of the joint FS-X program was actually a U.S. idea.[4]

Indeed, U.S. officials suggested a cooperative modification program only two months after receiving written confirmation that licensed production was a formal option under consideration by the Japanese government, nearly two years before a final cooperative deal on the FS-X actually emerged. Thus, from the earliest days of the FS-X discussions, the debate centered on the two options of indigenous development and cooperative modification, even though officially the Japanese were still actively considering the option of licensed production.

In hindsight, however, it could be argued that a more robust U.S. advocacy of licensed production might have produced positive results. Reagan administration officials grossly underestimated the inherent strength of the U.S. bargaining position in security matters with the senior Japanese political leadership in the face of rising economic frictions between the two countries. Furthermore, it is difficult to imagine that a more firmly held U.S. position in favor of licensed production would have caused any more antagonism and political disruption in the end than the course that was actually pursued.

Indeed, by mid-1987, it was clear that the Weinberger strategy of "pressuring without pressure" had failed completely, even for the option of cooperative development of a modified U.S. fighter. Ultimately, the United States resorted to heavy-handed and direct political pressure on the highest levels to force cooperative development of the FS-X based on a U.S. fighter. U.S. officials explicitly warned the Japanese political leadership that a decision in favor of indigenous development would threaten the very foundations of the U.S.-Japan security relationship. Mounting trade frictions, angry warnings from Congress, and the Toshiba Machine Tool scandal all weakened the ability of the Japanese political leadership to resist.

The same sort of pressures could have provided the backdrop for promoting licensed production at this time. The Japanese supporters of indigenous development were only marginally more opposed to licensed production than to cooperative modification. The political costs for the United States of taking an unbending position

---

[4]Indeed, some DoD officials even contemplated offering to develop an entirely new fighter cooperatively with the Japanese.

on licensed production could have hardly been much higher than they were for cooperative modification. Furthermore, the licensed production option would have eliminated the debilitating and drawn-out disputes over U.S. access to Japanese technology, particularly Mitsubishi's composite wing technology, which delayed negotiations for at least six months in 1988 and caused considerable bitterness on the Japanese side. Furthermore, a licensed production program would have prevented the endless maneuvering that took place later over the degree to which the aircraft would actually be modified. Clearly, licensed production would also have been a far more effective means of promoting DoD's broad objectives relating to Japanese equipment procurement and the U.S. industrial base. Finally, and most importantly, licensed production of a U.S. fighter would have provided Japanese industry with little of the R&D experience it so vigorously sought to increase its indigenous military R&D capabilities.

Some American officials have come to recognize the validity of this view. As one high-level Pentagon official deeply involved in the early FS-X negotiations later explained:

> It was the height of the Cold War. Back in 1985 and 1986 we were concerned that if we pushed licensed production too hard, the Japanese might just do nothing. We didn't want that. Our goal was to improve Japan's defense, not force them to buy U.S. fighters. We just didn't know how hard we could push them. It was not as apparent in 1986 as it is in 1992 how much leverage we have over them. The defense arena is not critical to the Japanese political leadership. They are willing to give in for the sake of the U.S. relationship. If we started over now, we would do it differently.[5]

Failure to Limit Modification of the F-16

Although Pentagon negotiators could have pressed harder for licensed production, DoD's proposal for cooperatively developing a modified U.S. fighter still could be seen as a reasonable compromise, permitting incorporation of at least some Japanese technologies to satisfy TRDI and Japanese industry, while maintaining good relations with the Japanese security establishment. A mini-

---

[5]Interview, senior Pentagon official, August 1992.

mally modified U.S. fighter could still meet virtually all U.S. economic and political objectives. GD and McDonnell-Douglas offered the Japanese such modification proposals. In many respects, this course could have represented the most reasonable compromise position for the U.S. government.

The problem with this approach emerged when the Pentagon failed to object sufficiently to the evolution of the modification proposals in 1986 and 1987 toward increasingly radical design changes incorporating extensive Japanese technological input. DoD's casual attitude stemmed in part from an assumption that, with a significant piece of the R&D work, U.S. contractors would dominate the design and development process for modifications based on their own Agile Falcon and Super Hornet modification design concepts. Some Pentagon officials apparently did not care how much the Japanese modified the baseline aircraft, as long as the program had significant U.S. government and industry involvement. It also appears that other Pentagon officials, and even GD managers, may not have been fully aware of the potential magnitude of the new development effort implied by such proposals as the SX-3 Plus, which grafted the most important Japanese technologies and design changes onto the basic U.S. Agile Falcon concept. The key events for the future of the program, however, were the acceptance of a Japanese firm as the lead contractor for the R&D phase and the allocation of final design configuration authority to JDA. Once these principles were firmly established early in the bilateral discussions, the Japanese essentially had won the ability to design and develop the aircraft in accordance with their own objectives.

Many officials in the Pentagon had originally hoped that joint FS-X development could help contribute to design concepts and technologies that would be directly applicable to U.S. defense programs. For example, at one stage in the discussions, some on the U.S. side believed that the Japanese could be convinced to accept the GD-developed Agile Falcon wing design for the FS-X. This wing, it was hoped, could be eventually retrofitted to U.S. Air Force F-16s as part of a major upgrade program. But since the U.S. side never developed a consensus on this or any other design and technology objective and did not contribute any funding to the cooperative R&D effort, America's ability to influence the technological evolution of the fighter remained minimal.

This problem was exacerbated by the program delays and diversion of attention to economic issues caused by the frontal attack launched on the deal by Congress in early 1989. While the American side argued bitterly over the consequences of transferring F-16 technology to a leading foreign economic competitor, the Japanese quietly completed the transformation of the planned R&D program into something much closer to an indigenous development effort than the Americans had ever intended. Once the dust had settled and R&D began in early 1990, the American firms found themselves stuck in the subordinate role of subcontractors in a development program dominated by Japan.

A key factor contributing to the success of the Japanese strategy was the fact that both U.S. industry and government officials in the mid-1980s tended to underestimate the military R&D capabilities of Japanese industry.

JAPANESE MILITARY R&D CAPABILITIES UNDERESTIMATED

U.S. Skepticism

From the beginning of the FS-X program, U.S. industry and government officials generally remained highly skeptical of Japanese claims about the advanced state of their fighter technology development and widely assumed that the U.S. side would retain overall control of any joint R&D effort with the Japanese, including fighter design configuration, technology applications, system integration, and technology transfer to the Japanese. This mistake led the U.S. side to pay insufficient attention to the Japanese strategy of transforming the FS-X into a virtually new fighter and contributed to the widespread U.S. lack of interest in Japanese military technology.

A primary reason for this view was the conviction that Japanese industry did not have anything approaching the experience and capabilities of U.S. contractors in overall fighter R&D and military subsystem development, especially given the cutting-edge technologies the Japanese wished to incorporate into the domestic FS-X. Officials felt that the vastly greater store of R&D experience and technological expertise possessed by U.S. industry would, in practice, place GD and other American firms in a dominant position in any joint R&D effort. The U.S. side would then control the design and technical evolution of the aircraft, making sure that modifications were either kept to a minimum or engineered primarily by the U.S. side to serve U.S. objectives.

In truth, the United States possessed inadequate knowledge about the technical capability of Japanese industry to build an indigenous fighter and the strength of its commitment to do so. In addition to nearly a total lack of detailed information about Japanese military subsystem R&D efforts, the conventional wisdom, seemingly confirmed by Defense Science Board Task Force studies and DoD technology assessment teams in the early 1980s, suggested that Japanese military R&D was severely underfunded and generally far behind technology developments in the commercial sector.

Although Japanese military R&D expenditures expanded rapidly in the late 1970s and early 1980s, their overall level remained relatively low. Between 1976 and 1990, military R&D grew on average by nearly 15 percent per year. But by the end of this period, these expenditures stood at a total of well under $1 billion, compared to $36.5 billion in U.S. military R&D outlays. At this time, total Japanese defense expenditures had grown to a level roughly equivalent to that of the United Kingdom, France, or West Germany (Alexander, 1993, pp. 5–6).[6] Although it was widely believed that official Japanese figures for military R&D significantly understated the true level of spending, it was still thought unlikely to be more than that spent by such countries as the United Kingdom and Germany (Chinworth, 1989). Yet these two European countries had long since given up the prospect of developing a first-line fighter on a national basis. The high cost of modern fighter development had driven them to pool their military R&D funds with Italy and Spain in the mid-1980s to provide the $12 billion over ten years thought necessary to develop the future European Fighter Aircraft. To U.S. officials at the time, the Japanese just did not seem to have either a history of military R&D spending to support the scale of outlays necessary to develop a cutting-edge indigenous fighter or the necessary R&D experience.

---

[6]Alexander uses an exchange rate of ¥202 rather than ¥143 to the dollar based on an assessment of actual purchasing power parity value. Even using the conventional exchange rate, Japanese military R&D expenditures remain well below $1 billion.

The Japanese Military R&D Strategy

Logical though the assumption was, it seems to have been unfounded, because in the 1970s and 1980s, the Japanese adopted a unique long-term incremental strategy to build up at least the basic level of experience and technological know-how necessary to provide them with a credible capability in the fighter development business. This strategy remained relatively low cost and largely hidden from public view, at least prior to the decision to launch full-scale indigenous development of the FS-X.

The Japanese strategy sought first to maximize the benefits to the domestic defense industrial base derived from licensed production of the U.S. F-15 and from other licensed production programs. Second, it called for utilizing other related military R&D programs to gain the skills necessary for the demanding task of fighter subsystem development and integration. Third, the strategy required TRDI to focus its growing but relatively limited funds on high-leverage military research programs that would directly contribute to enhancing key capabilities necessary to develop an advanced fighter and that could not be acquired from licensed production programs or from the commercial sector. Finally, Japanese industry, particularly the electronics sector, effectively mined the enormous expertise it had developed in civilian markets to "spin on" sophisticated commercial technologies and manufacturing techniques to key military subsystem applications.

Although foreign observers have often pointed out that system integration of major aerospace platforms and subsystems has historically been one of the most critical shortcomings of the Japanese aerospace industry, Japanese industry was well aware of its lack of experience in this area and took steps in the late 1970s to remedy it (see Samuels and Whipple, 1989, pp. 298–299; Alexander, 1993, pp. 41–42). Lacking the large-scale programs and huge financial resources available to the American defense industry, it adopted a relatively low-cost incremental strategy designed to help it acquire the system integration experience necessary for attempting fighter development. Specifically, it gained valuable experience through three major procurement programs launched in the mid-1980s: F-4EJ*kai*, XSH-60J, and XT-4 (see Samuels and Whipple, 1989, pp. 286–287 and 299–301).

In the F-4EJ*kai* program, Japanese industry took a basic re-inforced F-4 airframe and incorporated 47 new items. Since most of these items were proven systems of American design, Japanese industry could focus entirely on learning the complexities of integrating advanced avionics systems combined in a completely new way, a task denied them on the F-15 licensed production program. In the next program, the XSH-60J R&D effort, Japanese industry advanced a notch beyond the F-4EJ*kai* to the development and integration of indigenous avionics into the new but less demanding environment of a helicopter airframe. The government designated MHI as the prime contractor for system integration. MHI planned to complete development and begin flight testing in 1988, just as full-scale development of its indigenous FS-X was planned to begin.[7] The third effort, the MT-X (later XT-4) jet trainer program, offered industry an opportunity for additional experience with subsystem development and integration. Perhaps more important, this all-Japanese aircraft program served as a full dress rehearsal for designing, developing, and integrating an indigenous fighter.

While the F-4EJ*kai*, XSH-60J, and XT-4 programs furnished industry with the opportunity to advance its skills in system development and integration, these programs did not address other key areas of technology and subsystem R&D necessary for the future Japanese fighter. TRDI's integrated technology strategy met this challenge through support of a variety of dedicated military technology demonstration and development programs in the areas of advanced aerodynamic research, development of fly-by-wire (FBW) and control configured vehicle (CCV) technologies, advanced composite materials and structures, APA fire-control radars, and various avionics systems.[8]

Unlike their American and European counterparts, TRDI and Japanese industry had little experience with fighter technology demonstration programs. But in a major departure from the past, and one little noticed by the world aerospace community, Japan launched its own research into active control technology and un-

---

[7]Jackson (1985), p. 143; "Japanese Defense Budget Extends Growth Despite Strong Opposition" (1985), p. 72.

[8]"Japanese Near Decision on FS-X as Replacement for Mitsubishi F-1" (1986), pp. 87–88.

stable designs for future fighters in the late 1970s. TRDI selected the Mitsubishi T-2 in 1978 as a test bed aircraft to develop FBW capabilities and investigate CCV technologies.

In another area of great importance for future fighters, the development of composite materials and structures manufactured from carbon fiber-reinforced plastics, TRDI and industry pursued a clever strategy of combining dedicated military R&D, considerable spin-on from the commercial sector, and exploitation of capabilities acquired in both military and commercial collaborative programs with the United States. This effort later included development of radar-absorbing materials, an area critical for stealthy military aircraft.

The last and perhaps least well-known area of military technology for Japan's future fighter that TRDI focused on in the early 1980s was the development of an APA fire-control radar and a variety of key fighter avionics systems. Here, TRDI committed to a long-term dedicated military R&D effort for system development, working with MELCO (a major electronics firm that drew heavily on its strong commercial base in GaAs devices and other electronics technologies, experience gained from military licensed production, and possibly outside assistance with Western firms, such as Westinghouse Electronics). In addition, it appears that JDA and MELCO committed substantial resources and effort going back many years before the commencement of the fighter radar program in 1981 to investigating APA technology. According to Japanese press accounts, MELCO spent about ¥100 billion of its own money during the 1980s and early 1990s to develop the APA military radar technology.[9] This impressive effort, which would result in a full-scale engineering test model of an APA fighter radar ready for flight testing by early 1987,[10] clearly represents one of the most dramatic modern achievements by Japanese industry in advanced military technology. As a result, the much more experienced fighter developers in Europe have now fallen many years behind the Japanese in this critical fighter technology. Indeed, overall Japanese technological capabilities in this area are now generally

[9]*Nihon Keizai Shimbun,* January 26, 1993.

[10]"Mitsubishi Developing New Radar and Associated Weapons System" (1987), p. 6.

comparable to the most advanced developments on the American side.[11]

Thus, by the mid-1980s, Japanese industry and TRDI had made remarkable progress—at relatively low cost and low risk—in establishing the technological and organizational foundations necessary to support full-scale development of an indigenous fighter.

Given that the Japanese did have a credible overall capability to develop the Rising Sun fighter on a national basis—with the major exception of the engine—the assumption of U.S. officials that American industry, with its far greater expertise, could dominate configuration development and technology applications once the program was under way was unfounded.[12] Unfortunately, U.S. officials behaved as if this assumption was true. As a result, the Pentagon did not object to the evolution of the modification proposals in 1986 and 1987 toward increasingly radical design changes incorporating extensive Japanese technological input or to the establishment of MHI as the project's lead contractor—actions that inevitably drove the fighter toward Japanese indigenous development. Indeed, some officials seemed to feel that the more dramatic the modification proposal, the greater the practical need would be for the U.S. companies to provide the overall technology leadership and control of the R&D program. Yet once the R&D program was under way, GD and other American firms discovered to their chagrin that they were merely subcontractors carrying out the wishes of the Japanese lead contractors.

---

[11]The MELCO APA radar for the FS-X is actually less capable in a variety of performance areas than some existing U.S. fire-control radars based on more conventional technology. However, American radar experts have concluded that the FS-X radar demonstrates that Japanese engineers have mastered the basic technologies and skills necessary to produce a far more capable fire-control radar based on APA technology. This puts the Japanese far ahead of the leading European fighter developers in this key area of subsystem development.

[12]A former U.S. Air Force Program Manager for the FS-X characterized this question as follows (letter to the author, August 9, 1993):

> DoD negotiators understood the extent of the modifications contained in the top level [U.S. contractor design] proposals. What was not understood by the DoD or GD was that MHI planned to use the F-16 data as reference data rather than do an ECP [Engineering Change Proposal] to the F-16. MHI's goal was to develop a trained work force and do "their own thing." Working level JDA had the same goal. GD and DoD underestimated this. After the program started in March 1990 MHI made no attempt to develop a lightly modified F-16 and the U.S. had no authority—in the agreements—to temper MHI's "creativity."

But most importantly, the transformation of the FS-X was made possible because the U.S. government never clearly formulated and prioritized its objectives beyond the basic requirement for significant U.S. government and industry participation and never developed a coherent and unified strategy for attaining them.

CONFLICTING U.S. POLICY GOALS

Economic Versus Security Objectives

The success of the Japanese strategy of transforming a minimally modified F-16 into something approaching an all-new indigenous fighter can be attributed in large part to the fact that the U.S. side did not formulate and carry out a coordinated policy on weapon system collaboration with Japan that harmonized both U.S. military and economic objectives.

The problem arose from differing goals of the key U.S. players. Throughout the negotiations for an R&D MoU in 1988, and even more so during the bitter congressional and intergovernmental debates over FS-X in 1989, the important U.S. players pushed different agendas.

The Departments of Defense and State sought primarily to bolster the U.S.-Japan security relationship by discouraging a more autonomous Japanese defense industry capability and security policy by ensuring U.S. government and industry participation in the largest Japanese procurement program of the 1990s and by promoting ASDF equipment interoperability and integration with U.S. forces. However, Congress, the Department of Commerce, and other government agencies focused on chronic economic disputes with Japan and concerns about the long-term competitiveness of the U.S. aerospace industry. Here the main concerns were the trade deficit with Japan, the migration of American jobs overseas, and the decline of U.S. high-technology industries. Many in this camp considered the FS-X agreement a massive "giveaway" of advanced American aerospace technology to America's number one economic competitor. A basic assumption of this group was that F-16 technology could be applied by Japanese industry to commercial products. Indeed, many in this camp believed that Japanese industry sought collaboration primarily to acquire commercially useful American defense technologies. Finally, the U.S. prime contractors sought mainly to win the design competition and gain a piece of the FS-X program. The prime contractors had few concerns about how the second- and third-tier U.S. industry suppliers would fare in the deal and little motivation to limit Japanese modifications to the baseline F-16C design.

These various goals led to differing—and sometimes contradictory—strategies. The U.S. security establishment overwhelmingly focused on stopping Japanese indigenous development. The principal strategy was to win a substantial role for the U.S. government and industry in a joint R&D program. Many officials at the Pentagon also sought to minimize modifications to the baseline F-16/SX-3 design. Indeed, these officials had successfully urged GD in the summer of 1987 to withdraw another design proposal—the SX-4—because it entailed too much modification of the basic F-16. The Pentagon wanted a cost-effective fighter for ASDF that did not require a large and expensive new R&D effort. This meant keeping the modifications to a minimum.

However, Congress and the Department of Commerce sought to restrict the transfer of F-16 data to Japan, because they believed that this technology could be commercially beneficial to Japanese industry and would be applied to civilian products, especially commercial airliners. They also focused on maximizing workshare for U.S. industry, particularly during the production phase, which would provide considerable employment for U.S. aerospace workers. The Pentagon deemphasized the production phase, because it was much more concerned about influencing the design-and-development stage for strategic military reasons. Finally, Congress and others demanded greater reciprocity in the two-way flow of technology, insisting that Japan must flow back technology to the United States in partial compensation for the technology the U.S. transferred.

Some of these goals proved to be in direct conflict with one another. For example, to minimize modifications to the baseline F-16, the Pentagon sought to transfer all the necessary technical data packages to Japanese industry within the constraints of normal U.S. security procedures and national disclosure policies. After all, Japanese industry could not duplicate the baseline F-16 unless it had the plans and the technical data packages for the aircraft. Yet by seeking to restrict the transfer of U.S. data, Congress and the Department of Commerce played directly into the hands of the Japanese supporters of indigenous development. If denied American data, Japanese industry had no choice but to substitute indigenous development.

The case of the source code for the computer software that operates the flight control computer (FCC) for the F-16 digital FBW system is an excellent example of different U.S. policies working at cross purposes. Pentagon officials assumed that a "sanitized" version of the source code would be provided to Japanese industry and that GD would help develop the flight control system for the FS-X. They argued that Japanese industry would learn very little about the crucial developmental process. Japan would find out only the "know-how," not the "know-why," behind the development of the software. This would limit the capability of the Japanese to develop similar software on their own the next time around. Critics countered that the Japanese could use the F-16 source code to develop advanced flight-control technology that could be applied to commercial airliners. Furthermore, they did not believe Japanese industry could develop the source code entirely on its own. If the U.S. government denied access, critics claimed, Japanese industry would be forced to contract with an American company for the FCC software and buy it as a "black box" end item.

The critics won this fight. Yet instead of purchasing the software directly from the United States—as the critics had confidently predicted—Japanese industry went ahead and decided to develop this advanced technology entirely on its own. The Japanese government was willing to pay the greater cost of having the source code developed in Japan, because it knew that such an effort would greatly increase industry skills and capabilities. As a result, Japan is conducting an extensive new indigenous R&D program that will help catapult it into the forefront of advanced flight-control technology development. In all probability, Japanese industry would have learned far less about this demanding R&D process had it worked under GD supervision with "sanitized" F-16 FCC source code than it has developing its own source code from scratch.

Complex U.S. Oversight Structure Exacerbated the Problem

Unfortunately, the complex bureaucratic oversight structure established on the American side made reconciliation of these conflicting U.S. goals difficult as the program progressed through the R&D phase. That structure, which is illustrated in Figure 3, provides no clear lines of authority. Even long-time program officials have difficulty determining who is really in charge of overall U.S. policy on the program.

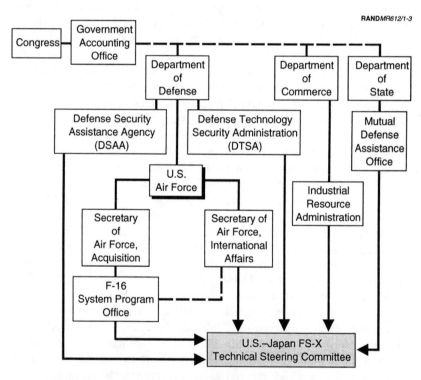

NOTE: This chart represents the program organization as of 1993. It has changed slightly since then.

Figure 3—Complex Oversight Structure Made Reconciliation of Conflicting Strategies Difficult

The U.S. Air Force, the formal executive agency for FS-X, is not the key player. With no commitment of Air Force budgetary resources to the effort, FS-X has been characterized as an "orphan program." The major players are the Department of Commerce—which, as the result of legislation passed in 1988, sits as a formal member on the FS-X Technical Steering Committee—the Pentagon's Defense Security Assistance Agency (DSAA), and the Defense Technology Security Administration. The latter two organizations often team up with the Department of Commerce to emphasize economic issues and questions of technology transfer and access to Japanese technology. This emphasis is further encouraged by close monitoring of the program by Congress through the Government Accounting Office (GAO).

The absence of a unified approach to negotiations—and the diffuse program structure that was a reflection of that disunity—help to explain why negotiations with Japan were so protracted. Pentagon officials were not prepared to begin serious negotiations when the Japanese launched discussions for the R&D MoU in late 1987. Even worse, a clear and consistent American strategy, which would reconcile the differing interests of the various players and establish U.S. priorities, had not been developed and, indeed, never was. Very little thought was devoted to formulating practical implementation measures and strategies to get them into the formal agreements. As a result, the U.S. government simply did not exhibit the unity of purpose or possess the necessary tools to decisively influence the direction of the R&D program. The Japanese side did not suffer from a similar disunity and confusion over goals. Their negotiators represented the key interests within the government committed to indigenous development of a Japanese national fighter. They knew what they wanted and how to get it.

The fact is that U.S. participants entered into negotiations with the Japanese before confronting the issues that would divide the policymaking community for many months after the terms of the negotiated agreement were made public in early 1988. Most important, the U.S. government allowed congressional opponents to prevail in shifting the terms of the debate from national security issues to dubious economic and technology transfer concerns—a shift that led to endless disputes over technology transfer and workshare, which diverted attention from Japanese industry's ongoing transformation of the FS-X.

MISGUIDED POLICY ON TECHNOLOGY TRANSFER AND FLOWBACK

As noted earlier, many members of Congress and the U.S. government believed that the transfer of FS-X technology would contribute to the emergence of an aggressive commercial aerospace industry in Japan that would repeat the successes of the Japanese auto and consumer electronics industries against their American competition. The argument is summed up in Clyde Prestowitz's influential 1989 *Washington Post* article, "Giving Japan a Handout," which helped ignite the debate in Congress over FS-X in 1989 (Prestowitz, 1989):

> [The FS-X deal] will transfer technology developed at great expense to U.S. taxpayers at very low cost to a country whose primary interest is not defense but catching up with America in aircraft and other high-technology industries . . . . [T]he United States could be creating a powerful competitor in its best export industry for a relative pittance in subcontract fees.

At the heart of Prestowitz's concern is the belief that the Japanese were more interested in leveraging U.S. military technology to achieve commercial gain, a view supported years earlier in a 1982 GAO study requested by Congress on the F-15 licensed-production program with Japan, entitled *U.S. Military Coproduction Programs Assist Japan in Developing Its Civil Aircraft Industry*. This widely read and influential report concluded that key Japanese objectives for entering into military licensed production programs were "obtaining advanced technology, enhancing their high-technology employment base," and "developing future export industries."[13]

Congress therefore sought to limit the transfer of F-16 technology to Japanese industry on the FS-X program. Yet this strategy was based on several false premises. First, the evidence overwhelmingly indicates that the F-16 technology transferred to Japan had little or no direct application to the commercial sector, even in commercial aerospace. Secondly, the strategy erroneously

---

[13]Comptroller General of the United States (1982), pp. ii and 4. For a typical press account, see "GAO Report Says Coproduction Pacts Aid Japan Industry" (1982).

assumed that Japanese industry did not have the experience or capabilities to substitute its own indigenous R&D for technologies and items denied by the United States and would therefore buy American items off the shelf. Thirdly, it represented a fundamental misunderstanding of the principal Japanese motives behind the FS-X program. Japanese program officials and industry participants were primarily interested not in gaining U.S. defense technology for commercial applications but rather in maintaining and further developing the Japanese national military R&D capabilities in aerospace weapon systems.

Commercial Spinoffs to Japan Exaggerated

The opponents of the FS-X deal in 1989 grossly exaggerated the potential commercial value of the F-16 technical data package for Japanese industry. Most aerospace industry experts would agree with the conclusion of one 1985 study that not much of the technology employed in fighter aircraft can be "readily transferred to applications in commercial aircraft." (Mowery and Rosenberg, 1985, p. 10.) Commercial airliners are large, relatively slow transport aircraft optimized for safety and low-cost, efficient operation. Fighters are small, densely packed aircraft optimized for high speed, maneuverability, and effective delivery of air-to-air and air-to-ground munitions. There are broad generic technologies and processes applicable to both types of aircraft. However, the performance and technological demands for developing modern fighters far exceed those for commercial aircraft in design, integration, materials, avionics, engines, and most other subsystems. In short, the relationship between developing fighters and airliners is roughly comparable to that between developing a high-performance sports car and a city bus.

In fact, the actions of the Ministry of International Trade and Industry (MITI) and of Japanese industry in the 1960s and later reveal a clear recognition of these differences. Development of military and commercial aircraft capabilities proceeded on separate tracks. Licensed production of the North American F-86 and Lockheed F-104 fighters and the further improvement of the Fuji T-1 jet fighter-trainer were the means by which Japanese industry sought to acquire technology and experience relating to high-performance fighter aircraft. But early on, MITI established a separate and much more direct avenue for civil aircraft development. The second Aircraft Promotion Law provided funding and established a special consortium of the leading aircraft companies—Mitsubishi, Kawasaki, Fuji, Showa Aircraft, Japan Aircraft, and Shin Meiwa—for the express purpose of developing a medium-sized commercial transport. The government financed over one-half the R&D costs. The resulting aircraft, the twin turboprop YS-11, proved to be a technological success but a commercial failure. Nonetheless, the YS-11 development effort provided Japanese industry with valuable experience much more directly related to developing future commercial transports than was work on the F-104 and T-1 (Mowery and Rosenberg, 1985, p. 10).

Interestingly, the Japanese actually moved from an indigenous civilian transport development program to a military program, rather than visa versa. In other words, instead of "spinning off" military technology to the commercial sector, they "spun on" commercial technology to the military sector. For example, the same consortium established to design and develop the YS-11 began design work on the C-1 medium military jet transport in 1966. Since then, Japanese industry has been far more adept at spinning on commercial technologies to defense programs than spinning off defense technologies to commercial efforts. Examples include the T/R modules on the FS-X APA radar and aspects of the composite materials technology used on the FS-X wing (see Chang, 1994).

As mentioned above, however, the development of Japan's commercial aerospace capabilities basically took a different tack from its military efforts. This is because Japanese industry understood that by far the most direct means of gaining experience with large commercial transports was to develop one. With the high expense and commercial failure of the YS-11, MITI's strategy shifted toward a policy of collaboration with leading U.S. manufacturers of airliners to further develop Japan's civil aircraft capabilities. Japanese firms were encouraged to establish subcontractor relationships with U.S. companies in the 1960s and 1970s for this purpose. With MITI guidance and support, Japanese firms formed a new consortium in 1973 and established a long-term collaborative relationship with Boeing. The consortium began joint design and development work with Boeing on the YX, which later became the B.767 airliner (Mowery and Rosenberg, 1985, pp. 10–11). In fact, the transfer of commercial aerospace technology and know-how during collaborative work with Boeing probably contributed to developing Japan's commercial capabilities. During congressional testimony in the early 1990s, Joel Johnson, Vice President of the Aerospace Industries Association, testified that such collaboration with Boeing "is far more relevant to the design and production of civil aircraft than any spinoffs from building fighters." (House, 1989, p. 168.)

Two recent U.S. government technical studies further support this view. In April 1990, the findings of a report undertaken by the nonpartisan Congressional Research Service (CRS) were presented to Congress and expressed strong skepticism about the alleged commercial usefulness of the U.S. technologies transferred to Japan in the FS-X effort. The CRS study focused specifically on the likely value for commercial aerospace applications of the F-16 technology transferred to Japan and concluded that "the F-16 airframe technology, the fighter design in general, has little potential for direct commercial spin-off." (Moteff, 1989, p. 8.) Even more interesting, a second study by the GAO, commissioned by Congress in 1991, explicitly set out to demonstrate that involvement by Japanese firms in the licensed production of the McDonnell-Douglas F-15 fighter in the 1980s provided those firms with the necessary capabilities to compete successfully against U.S. companies for major subcontracts on commercial transport programs conducted by Boeing and Douglas Aircraft. The study not only failed to prove its case, it tended to discredit it (GAO, 1992).

The GAO analysts carefully assessed the activities of 40 major Japanese aerospace contractors on the F-15 program to determine what participation they had on U.S. commercial aircraft programs. To their surprise, they discovered that the correlation was not very strong. Of the 40 Japanese F-15 companies examined, only 18 had any involvement in Boeing or Douglas airliner programs. Of these, only ten—25 percent of the total firms reviewed—provided closely related parts or components to both military and civil programs. Not surprisingly, the GAO was forced to conclude that "no single, causal relationship exists between Japanese companies' participation in the F-15 coproduction program and their involvement in the production and development of Boeing and Douglas civil airplanes." (GAO, 1992, pp. 3, 19.) Generally ignored by the GAO researchers, however, was the fact that a strong correlation did exist between suppliers of subsystems and components for the F-15 and similar items for the FS-X.

Although most experts agreed that the transfer of military technology to the Japanese did not help their commercial aviation industry, it clearly did help them in developing a fighter aircraft. Both the CRS study and the GAO concluded this, as did the overwhelming weight of the testimony from technical experts to Congress during the FS-X debates. This testimony supported the contention that it was not so much the specific F-16 technologies that the United States would transfer, but the experience the Japanese would gain in carrying out extensive modifications to the baseline F-16 design that would substantially increase their overall military aerospace capabilities, particularly in system integration.

Emphasis on Technology Flowback from Japan Was Political Symbolism

Like the issue of U.S. technology transfer to Japan, the issue of a reciprocal, two-way flow of technology predated the FS-X program. In the late 1970s, interest grew among some Pentagon officials in acquiring advanced Japanese dual-use technologies that could be used in the manufacture and development of new American weapon systems. This interest was motivated both by the emergence of Japan as a world technological leader in electronics and other fields and by the growing political necessity in the United States of balancing the flow of American defense technology to Japan with a reciprocal flow of Japanese technology to the United States.

This issue eventually became central on the FS-X program. The evidence suggests, however, that whatever the true value of Japanese technology, the great majority of U.S. government and industry leaders tended to discount it. The issue came to dominate the program primarily for political reasons. Unfortunately, the dominance of the issue in the negotiations proved to be highly counterproductive to U.S. strategic military interests.

Looking back on DoD technology initiatives that predated FS-X reveals several things. First, it shows that the legalistic, essentially ad hoc, and largely symbolic approach adopted by the Pentagon toward acquiring Japanese technology during the initial phases of the FS-X negotiations (1985–1988) directly mirrored the earlier, more general U.S. technology initiatives of the late 1970s and early 1980s. In both periods, the U.S. side first sought a broad legal framework for access to Japanese technology, while paying little attention to the formulation of realistic and practical mechanisms for its actual transfer. This approach suggests that the U.S. initiatives were sought largely for the political symbolism of Japan agreeing to the principle of greater reciprocity in the flow of technology. In both cases, Pentagon efforts to determine whether the American military services or U.S. defense contractors were actually interested in acquiring specific Japanese technologies appear to have been conceived as almost an afterthought.

Equally revealing is the extraordinary deficiency of knowledge on the U.S. side regarding specific Japanese technology developments, particularly in military R&D, as well as the general lack of interest in the DoD as a whole, the military services, and even among U.S. defense contractors in finding out more. This was part and parcel with the problem discussed earlier of underestimating Japanese military R&D capabilities. Conventional wisdom on the American side held that the Japanese had developed many interesting commercial technologies and manufacturing processes that could have military applications. U.S. officials, however, had little detailed knowledge about specific dual-use technologies that might be of interest and how precisely they could be applied to U.S. defense programs. With respect to military R&D, most American experts in government and industry believed Japan had little to offer the United States.

The lack of detailed U.S. knowledge of Japanese developments in defense-related technology was hardly due solely to shortcomings on the American side. Both during the early 1980s and later during the initial phases of the FS-X program, both the Japanese government and industry clearly resisted both sharing detailed technical information with the Pentagon and transferring dual-use commercial or military technology to the United States. A variety of domestic political and understandable commercial considerations primarily explain the Japanese reticence. However, there may have also been a conscious desire within segments of the Japanese military R&D establishment and industry to shield some of their more interesting defense-related technology developments from the prying eyes of the Pentagon and American contractors.

This is not to say that, in the mid-1980s, the Japanese definitely possessed a vast store of advanced military and dual-use technologies potentially of great value to American defense contractors. In fact, nobody in the United States really knows for sure one way or the other. In all probability, the conventional wisdom at the time at the Pentagon and in American industry was not that far from the truth. The point here is that few officials on the American side really knew what the truth was or cared to find out.

The bottom line for the FS-X program was that, although free and automatic flowback of derived technology continued to be a central DoD concern, that concern was primarily a symbol of technology reciprocity to appease Congress, not because there was a strong belief in the value and use of the technology.

Unfortunately, the need to appease Congress became less and less a symbolic gesture, especially with the release of a 1989 congressionally requested GAO study that raised serious questions about the value of Japanese technology to U.S. defense firms and U.S. military requirements for the technology. The study concluded, for example, that the "U.S. military requirement for the Japanese composite technology appears to be modest." (GAO, 1989, pp. 5–7.) While the study basically confirmed DoD's own beliefs, it also escalated congressional concern that the FS-X deal was not providing just compensation for the transfer of U.S. technology. Because Congress believed in the enormous commercial value of the F-16 technology expected to be transferred to Japan, the news that Japanese technology might be of little use and value drove concerns of a "technological giveaway."

As a result, FS-X supporters in industry and DoD were forced into the uncomfortable position of having to argue that the Japanese technology was definitely desired and potentially of great value, something they really did not yet know. Indeed, from very early on in the debate, GD pressed this point vigorously, even making the rather extravagant claim that the FS-X deal would permit access to "new Japanese technology *vital* to future military aircraft production."[14] For their part, administration officials played up the benefits of U.S. access. While emphasizing that DoD did not negotiate the FS-X agreement for the purpose of gaining access to

_____

[14]FS-X Gives U.S. 'Vital' Production Technology, GD Says," (1989). Emphasis added.

Japanese technology, Defense Secretary Dick Cheney and others nonetheless stressed the potential benefits to the U.S. defense industry of acquiring manufacturing technology for the APA radar T/R modules and the composite wing. Testifying to the Senate Committee on Foreign Affairs, Secretary Cheney insisted that the wing was "an area where we may have something fairly significant to learn from them." (U.S. Senate, 1989, p. 83.) In the area of T/R modules, Cheney claimed that access to Japanese manufacturing technology could potentially save the United States many millions of dollars on U.S. defense programs. Secretary of Commerce Robert Mosbacher concurred that "those are the two areas where we think we can have significant gain."[15] And DoD, industry, and the administration supporters of FS-X got some help from a 1989 CRS study that, in marked contrast to the GAO findings, concluded that the FS-X program provided U.S. access to Japanese "technology that could be potentially valuable," arguing that "purchase or possible licensed production of phased array radar transmitting/receiving elements may offer the most direct benefits." (Moteff, 1989, p. 11.)

In fact, MELCO's phased-array radar offers an example of one of the real concerns about the practical value of Japanese technology to U.S. defense industries. While DoD interest in the radar had always been publicly identified with Japanese production techniques for reducing the costs of manufacturing T/R modules and their associated GaAs MMIC chips, there was a nagging question of whether structural differences between U.S. and Japanese industry might seriously inhibit the effective transfer of Japanese technology. MELCO's manufacturing processes were based on a dual-use philosophy of simultaneously utilizing techniques and even the same machines—many of which are American made—developed in the commercial sector for both civil and defense applications. In contrast, much of the American defense industry was structured in a way that inhibited crossover between the commercial and defense arenas. In essence, the Japanese advantage in low-cost module production—if indeed it really existed—may have arisen more from differences in industry organization and structure, management philosophy, and procurement regu-

---

[15]Senate (1989), p. 83; also see "FSX Review Panel to Monitor Tech Transfer Compliance" (1989).

lations than from some identifiable manufacturing technology possessed exclusively by the Japanese. If this view proved correct, observers wondered how an entirely different R&D philosophy and industry structure, rather than a specific technology, could possibly be transferred from Japan to the United States.[16]

There may or may not be interesting process technology important to U.S. industry associated with MELCO's T/R modules. No one yet knows for sure. But what is certain is that most U.S. government and industry officials knew very little one way or the other about MELCO's process technologies when the FS-X program got under way and had virtually no idea how such technology could be effectively transferred to U.S. industry. The same could be said about Japanese composite materials technology. Recently, Vernon Lee of LFWC noted that the transfer of Japanese composite wing technology is one of the real success stories of the program. He added, however, that "just because you've got the technology, doesn't mean you need it." (Mecham, 1995.)

Even worse, the U.S. focus on technology flowback and access played a significant role in diverting U.S. attention away from the Japanese strategy of transforming the "minimally modified" F-16 into the Rising Sun fighter. Many disputes and problems arose in the course of implementing the technology access provisions of the agreements. Meanwhile, Japanese industry moved ahead with its plans to use the F-16 data as merely a baseline for experimenting with its own design excursions and technology applications. By 1993, the Japanese were generally in full compliance with the provisions regarding technology flowback and access. Yet, despite the great emphasis placed on this aspect of the program by the American side, the Pentagon, the services, and U.S. industry all resisted committing the resources necessary to evaluate the incoming

---

[16]Interviews, U.S. FS-X program officials. An extensive assessment of the development of the MELCO APA radar and the Japanese R&D strategy of spinning on technology from the commercial sector can be found in Chang (1994). Chang includes a detailed discussion of the structural differences between U.S. and Japanese industry that would make American adoption of the Japanese approach difficult. Press accounts reported early in 1995 that Westinghouse Electric had reached agreement with MELCO for purchase of the technology incorporated in the FS-X radar's high-power amplifiers. Ironically, Westinghouse reportedly wanted to use this technology exclusively for commercial applications, not for military programs. See "MELCO to Provide FSX Radar Technology to Westinghouse" (1995), p. 57.

Japanese technology fully. After the first three years of active R&D on the FS-X, there were still few indications that any Japanese technology would make a significant contribution to any existing or planned U.S. military programs.

In sum, then, the strong argument for the value of Japanese technology was pushed by supporters in 1989 largely to help counter the widespread criticism that the FS-X deal amounted to a massive giveaway of U.S aerospace technology to America's most feared economic competitor, even though neither side was truly convinced there was valuable technology out there. And even in the cases where there were identifiable interests in Japanese technology—such as the phased-array radar—there was legitimate concern in some quarters over whether it was actually transferable at all.

In the end, the United States may have ended up transferring military technology of little commercial value to Japan in return for access and flowback of Japanese technology in which U.S. industry was not really interested. The real beneficiary was Japanese industry, not because it received U.S. defense technology in some ill-conceived giveaway, but because it was able to use the program to develop and hone its own military R&D skills in preparation for the development of a future all-Japanese fighter.

Chapter Three

NEXT STEPS

Given this history of problems and difficulty, the question naturally arises: Should the U.S. government seek the continuation of the FS-X program into production? Our research suggests that the answer is a definite yes. Now that development of FS-X is nearly completed, full production of the aircraft is necessary to promote U.S. security and economic interests. Many of the potentially most important economic, technological, and political benefits of the overall program depend on the FS-X entering into series production.

BENEFITS OF PRODUCTION

The most obvious benefits are in income and jobs. Two-thirds or more of the total program revenue for U.S. industry is expected to be generated during the manufacturing phase. In addition, it has been estimated that FS-X production will provide nearly ten times as many man-years of employment for highly skilled U.S. aerospace workers as the development phase.

The production phase may also be crucial for the more effective transfer of interesting Japanese process technology to U.S. industry. As noted above, low-cost, high-yield Japanese manufacturing techniques for the composite wing box and the MELCO T/R modules have been the primary areas of Pentagon interest in FS-X technology since at least 1987. Most engineers would argue that process technology is best learned by doing. This is why GD and U.S. government negotiators fought so hard during the 1988 MoU negotiations to win the right to manufacture two of the six prototype wing sets at Fort Worth. However, it is not clear that the complete process for manufacturing the cocured composite wing during series production will fully mature during the R&D phase. Mitsubishi is still largely experimenting with tooling and manufacturing approaches during the wing development program. Undoubtedly the tooling and manufacturing processes will be refined considerably more during the actual production phase. Thus, full transfer of the process technology may require significant U.S. industry involvement in series production.

Finally, failure of the FS-X to enter production could wipe out the single most important political and military benefit of the program as originally conceived by the Pentagon: formal and extensive American involvement in the most important Japanese military procurement program of the 1990s. Production of the FS-X potentially guarantees an important U.S. role well into the next century, on both the government and industry levels, in ASDF procurement policies, as well as the overall evolution of the Japanese military aerospace industry. It can provide a unprecedented window on the future development of Japanese military technologies and capabilities. Continued joint FS-X program management potentially offers a unique forum for influencing Japanese policy, as well as for encouraging greater technological sharing and cooperation.

The currently planned production of FS-X, plus the possible future development of upgraded versions that are already being examined by TRDI and ASDF, could eliminate the rationale for launching an all-new indigenous fighter development program for years to come. As FS-X enters production in a post–Cold War environment of constrained defense budgets, it is likely to be increasingly viewed as a viable candidate for meeting other important ASDF replacement needs. The prospect of lengthening the production run and developing different versions of the aircraft will undercut those who advocate indigenous development of an all-new advanced trainer or even possibly an F-4EJ*kai* or an early F-15 replacement. As a result, the American government and U.S. industry could find themselves directly involved for years to come in a much broader array of major ASDF procurement programs than originally anticipated.

But the U.S. side may have to change its approach to the program to fully realize these potential benefits. A continued U.S. emphasis on technology transfer issues and access to Japanese technology could delay negotiations for the production phase and lead to a nominal production run or outright cancellation of the program. This would have adverse effects on U.S. security and economic interests, as discussed in detail below.

RISKS OF CANCELLATION

At the beginning of the original MoU negotiations in late 1987, the Pentagon had hoped to include an agreement on series production. However, in the course of the negotiations, DoD officials accepted the argument that the Japanese government could not be expected to commit formally to production until significant progress had been made in the development program. The first flight of the FS-X prototype, originally scheduled for 1993 and later slipped to 1995, was generally viewed as the critical milestone.

To address the American concerns over U.S. industry involvement during the production phase, the original agreements stipulated that FS-X production would not begin until the two sides negotiated a separate production MoU granting U.S. contractors a share of the work roughly comparable to their workshare during R&D. As an additional assurance, engine manufacturing data would not be released to Japanese industry until after an acceptable production MoU was signed. However, the lack of an explicit guarantee of the same 40-percent workshare for U.S. contractors during production that was specified for the R&D phase soon emerged as one of the main focal points of criticism of the FS-X deal during the congressional blowup in early 1989. As a result, the Japanese ultimately accepted new side agreements during the "clarification" process that specifically guaranteed that U.S. industry would receive 40 percent of the work during production.

Nonetheless, some U.S. skeptics remained dissatisfied with the provisions dealing with FS-X production. Their criticism centered on the failure of the accords to require Japan to enter into production. They pointed out that nothing in the MoU, the side letters, or other program agreements committed the Japanese government to manufacturing and procuring the FS-X once development is completed.[1] They argued that the Japanese could use the FS-X R&D program as a relatively low-risk dry run for full-scale indigenous

---

[1]However, Japan is obligated to pay cancellation fees to the U.S. government and industry if the program does not enter the production phase.

development at a later date. With the F-16 technical data package in hand and the assistance of seasoned U.S. contractors, Japanese firms could gain invaluable experience in the demanding task of fighter development, integration, and testing. Upon completion of R&D, the Japanese government would cancel the production phase, citing a reduced threat, escalating costs, technological problems, or friction over technology transfer. Soon thereafter, the skeptics warned, Japan would launch an ambitious indigenous fighter program making use of the experience gained during FS-X R&D—unencumbered by American participation and constraints.

The majority of U.S. program officials have always considered this unpleasant scenario to be rather far-fetched and paranoid. They point out that replacement of the aging Mitsubishi F-1 support fighter, already long delayed, must begin at the end of the 1990s. This replacement schedule does not leave sufficient time to develop a new indigenous fighter. Furthermore, with F-15 and P-3 licensed production ending in the latter half of the decade, Japanese industry will need a major new production program at that time to keep their factories and workers employed. Perhaps most importantly, U.S. officials insist that Japan's leadership would not risk the political breach with the United States that would likely result from a decision not to go into production.

While these arguments are compelling, a Japanese decision to forgo production of the FS-X is hardly inconceivable, particularly in view of the growing downward pressure on the Japanese defense budget following the collapse of the Soviet Union. Much depends on the final outcome of the R&D program in terms of cost and aircraft performance, as well as the continuing evolution of the cooperative arrangements with the U.S. side, particularly in the area of technology transfer. While the government has already authorized nearly all the funding for the formal R&D phase, years of prototype flight testing and other developmental tasks will still have to be financed. No one knows how well the aircraft will perform once flight testing has begun. The all-composite wing, the complex new avionics, the flight-control system, and many other technological aspects of the aircraft are still considered to be areas of relatively high risk for Japanese industry where unforeseen technical problems may still arise. Friction with the United States over the technology flowback and stepped-up efforts to gain access to Japanese technology may also undermine the program. Finally, negotiations over the production MoU itself are likely to prove difficult and may diminish Japanese support for the production phase.

Any possibility, no matter how remote, that production may not take place should be cause for considerable concern on the U.S. side. While outright cancellation may be unlikely, a small nominal production run of 40 to 60 aircraft is a real possibility. With cancellation or a limited FS-X production program, Japanese industry would be presented with a variety of attractive new options for indigenous development free of any direct American involvement.

TRDI has already funded considerable research on a future fighter aircraft. Indeed, by late 1994, various press accounts had revealed that TRDI was requesting an initial ¥1 billion for FY 1996 to launch development of a all-new Japanese advanced stealth fighter, called either the FI-X or FD-X. A prototype technology demonstrator is envisioned to fly in 2007. Reportedly, JDA kicked off the development effort by allocating funds to Ishikawa-jima Harima Heavy Industries in the FY 1995 defense budget to begin R&D on Japan's first fighter turbofan jet engine, intended for the FI-X. The new Japanese fighter would incorporate cutting-edge technology and subsystems, such as a conformal radar, thrust-vectoring engine nozzles, and considerable stealth technology. According to press accounts, the FI-X would draw extensively on experience gained in the FS-X program, particularly in the areas of cocured composite structures, stealthy radar-absorbing materials, FBW flight controls, phased-array radar, and so forth. Unlike the FS-X, however, the FI-X is intended to be an all-Japanese program, even for the engine technology, in which Japanese industry has traditionally lagged greatly behind the United States. JDA planners see FI-X as a potential replacement for ASDF's top-of-the-line McDonnell-Douglas F-15Js. FI-X could evolve into a direct competitor with America's newest fighter, the stealthy F-22 under development by Lockheed and Boeing. This has caused concerns for some observers, which are summed up by one U.S. expert as follows:

It's easy to imagine a future where Japan is building high-tech fighter aircraft more capable than anything we have the funding to produce.[2]

Thus, the bulk of the potential economic, technological, and political-military benefits for the United States from the FS-X program may depend on ensuring that this fighter enters into full series production. Most of the income and jobs will come from the production phase. Full transfer of Japanese process technology may require extensive U.S. industry involvement in production. Perhaps more important, FS-X production may help reduce the incentives for development of an all-new Japanese fighter for many years to come and may permit the United States to remain fully engaged in the evolution of the Japanese military aerospace industry. But these benefits will not automatically arise if Japan decides to go ahead with production. Rather, their realization will in large measure depend on the specific content of the production MoU that the two sides negotiate.

HOW TO DO BETTER

The United States needs to make the FS-X program work better and ensure that full series production takes place. How can this be done? The U.S. government should develop a carefully thought-out and well-coordinated high-level strategy to guide negotiations over the content of a production MoU. Without careful planning and preparation, the U.S. side risks further disruptions and disappointments on the program.

U.S. program officials spent countless frustrating hours in the early stages of the R&D program debating with their Japanese counterparts the precise meaning of specific words or phrases in the original agreements. Entire new documents had to be negotiated just to clarify various aspects of the original MoU. Therefore, the U.S. side would be well advised to enter into the new negotiations with a clear understanding of its objectives and priorities and to make sure that they are explicitly spelled out in the MoU. Im-

---

[2]Natalie Golding, British-American Security Information Council, quoted in Towle (1993), p. 9.

portant words and phrases should be defined with great care and precision.

Our research indicates that the United States should also consider reducing the emphasis placed in the past on the legalistic aspects of technology flowback and access and should adopt a flexible approach to the question of workshare percentages during production and the allocation of specific work tasks.

Continuing disputes over technology flowback and access could seriously delay the negotiations for a production MoU. The American side needs to stand back and to review this question seriously from the technological and the political perspectives. It should review the potential costs and benefits, as well as the practical feasibility of gaining real benefit from Japanese technology through the mechanisms established in this program. The U.S. government should seek to determine what U.S. companies—if any—are seriously interested in the Japanese technologies that might be made available in the future through the program. The challenge to the American side will be to determine a way to resolve legalistic disputes over technology access without causing political disruption to the program, while enhancing the prospects for meaningful access to interesting Japanese technologies.

Another cluster of problems involves the question of achieving the mandated 40 percent of the production budget for American industry and the actual division of specific work tasks during production. The side letters negotiated by the Bush administration during the FS-X clarification process guarantee U.S. industry a 40-percent share of production work. Achieving this will undoubtedly prove to be one of the most politically challenging and sensitive aspects of the negotiations. The problem of assigning specific work tasks during the production phase may be difficult to resolve. Because of a variety of factors, it is likely that negotiators will be forced to divvy up work tasks for the production phase somewhat differently than was done during R&D, which could cause significant difficulties.

This is an additional reason that both government and industry need to devote greater resources to assessing Japanese derived and nonderived data. The negotiations over production work division may be difficult. The Japanese will know exactly what they want to get out of the negotiations. The U.S. side should enter the negotiations with a clear idea of what tasks the United States would like to be allocated and why. If careful examination of Japanese data indicates that there may be manufacturing processes or technologies of genuine interest, the American side should consider targeting these areas for production to ensure their effective transfer.[3]

The bottom line for U.S. negotiators, however, should be to remain focused on maintaining and encouraging continued U.S.-Japan procurement collaboration for ASDF's next fighter. The American side must avoid at all costs the fundamental mistake of permitting legalistic disputes over abstract rights of U.S. access to Japanese technology—rights that may never be fully exercised—to undermine the continued survival and full series production of the FS-X. If FS-X is canceled or only produced in relatively small numbers, it is only a matter of time before the all-Japanese FI-X or some other indigenous fighter takes its place.

---

[3]There is nothing in the existing FS-X agreements that prohibits the U.S. side from seeking participation in the manufacture of any part of the aircraft, no matter who led development during the R&D phase. Thus the requirement for a 40-percent U.S. share could in principle be used as leverage to gain access to manufacturing techniques in any area of interest. For example, it appears doubtful that MELCO's manufacturing processes for the T/R modules, so highly touted during the 1989 debates, will ever be fully transferred unless a U.S. company takes part in their manufacture. Tens of thousands of T/R modules will be needed during the production phase. If the process technology really appears interesting to U.S. experts, production of some or all of the T/R modules under license in the United States could be sought.

Chapter Four

LESSONS FOR FUTURE COLLABORATION

DEFENSE TECHNOLOGY COOPERATION SHOULD BE VOLUNTARY

The first lesson of the FS-X experience is simple: Joint R&D ventures are likely to experience difficulties if both sides do not perceive real technological, economic, and political benefits in the program. An effective two-way transfer of technology requires active and voluntary participation initiated by both sides, motivated by each partner's belief that collaboration will result in a significant net technological gain. Such a situation arises when both participants can make technological and financial contributions to the joint effort that complement each other and directly assist each side in achieving its own objectives.

Development of the X-31 fighter technology demonstrator is an example of such a program that is worth briefly reviewing.[1] It is particularly interesting because it represents the first example of true *ab initio* international codevelopment of a military aircraft involving the United States. The X-31 is a fighterlike test aircraft developed to explore the enabling technologies and operational utility of radical improvements in fighter maneuverability. Unlike the FS-X, the X-31 is only a technology demonstrator and will never be fully developed and series-produced as an operational

---

[1]This account is based on unpublished research conducted by the author in 1992, which included extensive interviews with key government and industry officials on both sides of the Atlantic. Competent overviews of the program can be found in Lerner (1991), pp. 29–37; Wanstall and Wilson (1990), pp. 405–407; and "X-31: The Wonder Plane" (1990), pp. 80–83.

weapon system. Nonetheless, the technological and organizational challenges encountered in the design, development, manufacture, and flight testing of the two X-31 prototype aircraft in many respects parallel those encountered on a typical fighter R&D program. Indeed, designers consciously patterned the X-31 configuration on a serious design concept for a future European combat fighter.[2]

The X-31 aircraft was developed and manufactured collaboratively in the late 1980s by Rockwell International in the United States and Messerschmitt-Bölkow-Blohm (MBB) in Germany, now part of Deutsche Aerospace. The program is sponsored and funded by the U.S. Advanced Research Projects Agency and the German Ministry of Defense. On the American side, the U.S. Navy acts as the executive authority, while the U.S Air Force and NASA cooperate closely with the program.

The X-31 required the design, development, and integration of a variety of advanced technologies and subsystems into a unique aerodynamic configuration that provided highly unorthodox maneuvering capabilities for use during air combat. Among the most important technological challenges during X-31 development were the overall aerodynamic design configuration, the remarkably complex flight-control system, and the novel thrust-vectoring system employing carbon-carbon composite paddles attached to the tail pipe. It is the unanimous view of program officials and technical experts on both sides of the Atlantic that the R&D program generated a substantial two-way flow of technology and expertise. All the U.S. program managers involved in the development of the aircraft believe the Germans transferred technology, data, and know-how equal to or greater than that transferred from the

---

[2]In the late 1970s, German industry developed a design concept based on a delta-canard configuration called the TKF (*Taktisches Kampflugzeug*, or tactical combat aircraft) as a candidate for a future collaborative European fighter. While the general TKF design configuration eventually served as the basis for the European Fighter Aircraft, the European governments rejected the German industry requirement for including unorthodox maneuvering capabilities. The Germans then sought to develop a prototype jointly with the Americans patterned after the TKF design to demonstrate the technological and operational feasibility of supermaneuverability. This effort led to the initiation of the X-31 program.

United States.[3] Yet the R&D program encountered few major problems and virtually no disputes involving technology transfer from either side.

Program officials on both sides agree that the strong perception of mutual technological benefit, particularly on the industry level, was the key to promoting successful technology reciprocity. Both parties brought substantial technical data and R&D experience to the X-31 program from prior national programs that were complementary and freely shared it. MBB had conducted years of independent wind tunnel tests and simulation studies on various aerodynamic configurations.[4] Most importantly, the Germans had developed the basic design concept for supermaneuverability on which the X-31 would be based and offered it to the Americans. Rockwell had also conducted considerable R&D on novel configurations for enhanced fighter maneuverability, including work with an unmanned flying technology demonstrator, the highly maneuverable technology (HiMAT) vehicle.

On their own initiative, the two firms undertook collaborative exploratory research from 1981 through 1984, financed with corporate funds. After gaining interest in their novel concepts from elements within the U.S. Air Force R&D community and elsewhere, the two companies successfully sought funding from their respective governments in 1985 for a joint feasibility study. In June 1986, U.S. and German government officials signed a Memorandum of Agreement (MoA) for the cooperative funding and development of the X-31.

The remarkably brief and simple MoA calls for "a fair and cooperative research, design, and flight test program of [X-31] technologies." Indeed, the hallmark of the X-31 program was collaboration on virtually all key aspects of the R&D effort, including the maximum feasible sharing of the resulting data within the normal constraints of each country's national disclosure policies. As an example, the primary technical challenge during the initial phase of

---

[3]Based on multiple interviews with Col John Nix (USAF), Lt Col Michael Francis (USAF), John Retelle, and James Allburn, former X-31 program managers, in 1991 and 1992.

[4]Additional studies were conducted collaboratively with McDonnell-Douglas in the late 1970s.

R&D was development of the basic X-31 configuration. Rockwell and MBB split the total effort that went into configuration development almost equally, as measured in engineering man hours. The basic wing configuration and leading-edge sweep derived from MBB's extensive database on its J-90/P-30 fighter concept, but Rockwell developed the detailed shape of the airfoil involving wing twist, camber, thickness, and so forth. On the digital FBW flight control system—a system far more complex and challenging than that required for the FS-X—the Germans generated the basic control laws, an American subcontractor wrote the code, a U.S. vendor supplied the computer, and Rockwell and MBB integrated and refined the overall system in close collaboration.[5]

The experience with the X-31 R&D program dramatically contrasts with the history of the FS-X program. Both X-31 partners brought their own complementary technology, data, and expertise to the program and worked closely together on virtually all aspects of the full-scale development phase. The FS-X participants, on the other hand, quickly became mired in endless disputes over technology transfer and access, with U.S. industry essentially shut out of many of the most important areas of design and development. Through the X-31 program, the United States is gaining a potentially valuable new complex of military technologies and operational concepts, much of it based on German contributions. While the Japanese are now transferring much data from the FS-X program to the U.S. side, the ultimate value and benefit of that data remain uncertain.

The most fundamental difference between the two programs can be found in the basic interests and motivations of the participants, especially on the industry level. The X-31 started solely at the initiative of two companies that saw economic and technological benefit in working together and sharing their technological know-how to advance their common objectives. Joint government funding was won only later, when officials in the military R&D establishments and armed services on both sides of the Atlantic were convinced of the potential military benefits of exploring the new technologies jointly.

---

[5]Based on multiple interviews with M.R. Robinson, Director, Advanced Systems & Technology Development Programs, Rockwell International; and Hannes Ross, Manager, Advanced Design, MBB-Deutsche Aerospace, in 1992.

In the case of FS-X, Japanese industry and most of the military R&D establishment vehemently opposed collaboration with the United States. They saw nothing of great benefit that they would gain through U.S. participation, but much that they could lose. Many Japanese suspected that the U.S. government sought only to suppress the further development of their domestic military aerospace industry and to gain unfair access to their own commercially valuable technology. ASDF was also generally not enthusiastic about joint development. For its part, American industry supported collaboration, but primarily as a means of winning participation in a potentially lucrative program from which it otherwise would be excluded. U.S. companies did not generally believe the Japanese side had a great deal to offer in the way of new technologies or know-how, with the possible exception of the cocured wing. At most, they hoped that Japanese money could be used to help them develop versions of existing aircraft that later might be sold to the U.S. services or other allies. There are few indications that, in the early stages of the program, the U.S. Air Force or any of the other services viewed FS-X as an important means of acquiring significant new technologies.

The U.S. government, rather than industry or the services, led the effort to guarantee access to Japanese technology through the FS-X program. Yet this effort had a major symbolic political component. Many Pentagon officials were concerned about the mounting criticism from Congress and elsewhere that past licensed-production programs represented a one-way transfer of advanced aerospace technology to America's economic competitors. Greater technology reciprocity on FS-X could help counter these criticisms. Yet in the early phases of the program, government officials expended relatively little effort identifying specific Japanese technologies that might be of interest to U.S. industry or the services. The emphasis on the MELCO APA radar and the cocured wing arose relatively late in the negotiating process, in part as a response to congressional criticism over the lack of technology reciprocity. U.S. officials initially targeted the radar in part because it was the only Japanese avionics system planned for the FS-X about which they knew some details. Neither did government officials carefully think through the details of a realistic and practical mechanism through which Japanese technology could be transferred and successfully applied to U.S.

programs until the program was well under way. Once the final deal was sealed, implementation of the program was handed over to the U.S. Air Force, with little specific guidance on how to make it work.

Without a strong confluence of perceived interests, particularly among the participating industries and military R&D establishments on both sides, the mutually beneficial sharing of technology and R&D expertise is difficult to implement. While American officials should continue to pursue every reasonable avenue for gaining access to potentially useful Japanese technologies through the FS-X program, they should also seek to structure future programs in ways that will more naturally promote a mutually beneficial two-way flow of technology. More effective technology sharing would emerge from programs that involve the following:

- Foreign partners who possess technology or data that is of clear potential importance to American weapon development and who are willing to share it
- Industry partners who actively seek collaboration and offer complementary technological strengths and contributions
- Genuine interest from a military service on each side in developing the technology or procuring the resulting weapon system
- Genuinely collaborative R&D
- Equity in technology access and restrictions
- Financial contributions from all participating governments equal to workshare.

Unfortunately, there are signs that the U.S. government has failed to learn the lessons of the FS-X experience. A recent Pentagon initiative calls for joint procurement with Japan of a Theater Missile Defense (TMD) system. As reported in the press, the Pentagon initially offered to transfer advanced aerospace technology to Japan for the joint development of a TMD system aimed at protecting Japan against the North Korean ballistic missile threat. In return, Japan would transfer or permit access to many of its dual-use technologies. However, there was no indication that the DoD either had identified specific Japanese technologies of clear interest to U.S. industry and the services or had developed a realistic approach to how this technology could be usefully transferred.[6]

It could be argued that, because of domestic politics, Japanese companies can never be expected voluntarily to seek R&D and technology collaboration with American defense contractors. Yet U.S. officials may feel compelled by American politics to demand technology reciprocity from Japan when encouraging Japan to procure a weapon system collaboratively with the United States for political and military reasons. Nonetheless, forced collaboration does not seem to promote fruitful technology exchanges. Some other mechanism of satisfying U.S. domestic economic concerns while promoting military and political objectives through defense equipment programs with Japan must be sought.

The American government should avoid disrupting the U.S.-Japan security relationship again over politically symbolic issues of technology reciprocity when no obvious benefit to American industry may exist. No one wants the TMD initiative to turn into another FS-X. Can we afford to repeat the same mistakes?

CODEVELOPMENT PROLIFERATES MILITARY R&D CAPABILITIES

A second lesson of the FS-X experience is that cooperative military development programs carry the potential for significantly aiding a foreign country that is trying to increase its independent military R&D capabilities. In the long run, such programs can lead to a reduction of U.S. influence over the security policies of important allies and can help establish competitive foreign defense industries that may undermine the U.S. defense industrial base. Licensed production, on the other hand, usually transfers little technology of significant commercial value to advanced industrial countries like Japan and does little to promote the design and development know-how necessary to develop modern weapon systems. If codevelopment is the only alternative, it must be struc-

---

[6]Licensed production of a U.S. system is reportedly one option also being offered to the Japanese. See Sanger (1993), p. 5, and Jameson (1993), p. A8. DoD officials later modified this initiative by separating the technology reciprocity aspects into a separate initiative called Technology for Technology.

tured and managed very carefully from the U.S. side to minimize these risks.

Particular attention needs to be focused on the question of the transfer of expertise—as opposed to technology—during cooperative military R&D programs. Despite the trend toward globalization in high technology, the United States still possesses significant leads in most military technologies and, more importantly, in the formulation of requirements and the design and integration of sophisticated weapon systems. Codevelopment programs can proliferate the specialized skills built up by America's leading defense contractors throughout the decades when R&D was conducted on a scale far beyond what any other nation could afford.

Much of the basic rationale underlying the Pentagon's policy in the mid-1980s toward the FS-X program remains valid. U.S. political, military, and economic interests are generally not well served by the global proliferation of the technological and industrial capabilities to develop advanced weapon systems independently. This is particularly true in the case of Japan. A continuing buildup of the Japanese defense industrial base while the U.S. draws down its forces in the Pacific could encourage a more autonomous Japanese security posture. It could also fuel regional arms races and promote instability, as South Korea, China, and other neighbors seek to counter newly acquired Japanese capabilities.

An expanding Japanese defense sector may also pose a potential threat to the long-term health of the U.S. defense industrial base. The export of military equipment by Japan is prohibited only by Cabinet policy, not by legislation or the constitution. Beginning in the early 1980s, leading Japanese industrialists called for modification of the Cabinet ban on military exports. Because of the limited domestic Japanese market for military hardware and the high costs of military R&D, the development of indigenous systems greatly increases pressures for export. The high quality of many of Japan's defense technologies, particularly those "spun on" from the civilian electronics and other commercial sectors, could represent a major competitive challenge on the world market at a time when U.S. defense contractors may become increasingly dependent on foreign sales.

For these reasons, DoD officials in the Reagan administration were justified in seeking to discourage indigenous development of the FS-X. And they did not unfairly single out Japan. Throughout the 1980s, they launched several major efforts to convince the United Kingdom, Germany, Italy, and Spain not to develop the European Fighter Aircraft. They tried unsuccessfully to undermine French resolve to develop the Rafale fighter. After a brutal political battle, they finally forced the Israelis to cancel development of their Lavi fighter. They generally refrained from attacking the Swedish Gripen and the Taiwanese Indigenous Defense Fighter, because these are smaller, less-capable aircraft, and because U.S. industry involvement was already massive.

Yet while the Pentagon's ultimate objectives were justifiable from the American perspective, its strategy for implementation was seriously flawed. For a variety of reasons already discussed, the U.S. side lost control of the technological evolution of the FS-X, permitting Japanese industry to modify the basic F-16 design far more than originally anticipated.

One of the most compelling reasons U.S. industry has moved toward licensed production and cooperative development programs that transfer the industry's own expertise and technology is the fear that, if the United States refuses to cooperate, foreign countries will turn to European or other producers for collaborative deals. This argument was used to good effect by the Japanese on FS-X, the Koreans on the Korean Fighter Program (KFP), and many other allies. Yet this argument underestimates the political importance of the larger security relationships the United States maintains with allies and the generally superior quality of U.S. weapon systems. With the end of the Cold War, U.S. policymakers should consider more directly and forcefully linking the overall benefits of its security relationships with the need for allies to purchase major American weapon systems. For example, the South Koreans could threaten to collaborate with the French to develop or license-produce a new fighter if the United States restricts industrial offsets. But how credible is this threat? Would the South Korean government feel confident of French military support if the North Koreans invade? Would South Korean pilots perform better against North Korean MiGs flying French Mirages or U.S. F-16s or F-18s?

Thus, policymakers need to assess important questions about the proliferation of weapon development capabilities and about economic issues when evaluating cooperative weapon system development programs. U.S. officials might consider easing restrictions on arms exports and tightening controls over cooperative development efforts. Selling even the most advanced and sophisticated weapon systems to allies retains far more control in the long run for the United States over technology proliferation than co-developing a somewhat less capable system that helps move the foreign partner closer to industrial and technological independence in advanced weapon system development. The United States is still the world's leader in defense technology and military R&D. With declining defense budgets, few new major military R&D programs on the horizon, and a dramatically shrinking defense industry, how many years will that leadership position be maintained?

Yet, as demonstrated by the X-31 program, cooperating with technologically advanced allies on basic military R&D in narrowly defined areas can provide significant benefits to both sides. This is particularly true if the United States can clearly identify specific foreign technologies or data that would genuinely contribute to American weapon development and that could be made available to U.S. industry through collaboration. Currently, the United States is pursuing several collaborative programs with Japan under the auspices of the Science and Technology Forum aimed at conducting basic research in specific military technologies.[7] Some of these initiatives may turn out to be far better models for a more effective type of military technology collaboration with Japan than the FS-X or the proposed TMD initiative.

The FS-X program and other collaboration ventures with Japan can still be shaped to serve the best long-term military and economic interests of the United States and Japan. They can provide U.S. involvement in and influence over the development of the defense industrial sector in Japan for many years to come. They could provide an opportunity for the U.S. defense industry to learn from the successes of Japanese defense-industry management and structure and ultimately could contribute to the emergence of more genuinely collaborative and mutually beneficial military R&D between the two countries. They can help cement a stronger military and security relationship with one of America's most important al-

---

[7]The technologies include millimeter-wave infrared dual-mode seekers, advanced steel manufacturing, ship demagnetization, and ceramic materials for rocket engines and fighting vehicle propulsion systems. See Opall and Usui (1993), p. 3.

lies in the post–Cold War era. But they could also erupt again into more debilitating and destructive disputes between the two partners over emotional technology-transfer and trade issues.

Much depends on the planning and foresight of the American negotiators, the ability of Congress to assess military technology collaboration programs in an unemotional and rational manner, and the response of the Japanese government.

REFERENCES

Alexander, Arthur J., *Of Tanks and Toyotas: An Assessment of Japan's Defense Industry*, Santa Monica, Calif.: RAND, N-3542-AF, 1993.

Chang, Ike Y., Jr., *Technology Access from the FS-X Radar Program: Lessons for Technology Transfer and U.S. Acquisition Policy*, Santa Monica, Calif.: RAND, MR-432-AF, 1994.

Chinworth, Michael W., *Financing Japan's Defense Buildup*, The MIT Japan Program, Center for International Studies, Boston: Massachusetts Institute of Technology, 1989.

Comptroller General of the United States, *U.S. Military Coproduction Programs Assist Japan in Developing Its Civil Aircraft Industry*, ID-82-23, March 18, 1982.

"FS-X Gives U.S. 'Vital' Production Technology, GD Says," *Aerospace Daily*, February 3, 1989.

"FSX Review Panel to Monitor Tech Transfer Compliance," *Aerospace Daily*, May 4, 1989.

GAO—See U.S. General Accounting Office.

"GAO Report Says Coproduction Pacts Aid Japan Industry," *Aviation Week and Space Technology*, March 29, 1982, Vol. 116, No. 13.

Jackson, Paul, "Reluctant Samurai, Part II: The Maritime and Ground Self-Defense Forces," *Air International*, September 1985.

Jameson, Sam, "Japan Ready to Discuss Missile Defenses," *Los Angeles Times*, November 3, 1993.

"Japanese Defense Budget Extends Growth Despite Strong Opposition," *Aviation Week and Space Technology*, March 18, 1985.

"Japanese Near Decision on FS-X as Replacement for Mitsubishi F-1," *Aviation Week and Space Technology*, March 10, 1986.

Lerner, Preston, "Stall Tactics," *Air & Space*, April/May 1991.

Lorell, Mark A., *Troubled Partnership: A History of U.S.-Japan Collaboration on the FS-X Fighter*, Santa Monica, Calif.: RAND, MR-612/2-AF, 1995.

Mecham, Michael, "Japan's FS-X Fights Costs," *Aviation Week and Space Technology*, January 23, 1995.

"MELCO to Provide FSX Radar Technology to Westinghouse," Foreign Broadcast Information Service (FBIS), JPRS-JST-95-015, March 13, 1995.

"Mitsubishi Developing New Radar and Associated Weapons System," *Comline Transportation*, August 3, 1987.

Moteff, John T., *CRS Report for Congress—FSX Technology: Its Relative Utility to the United States and Japanese Aerospace Industries*, Congressional Research Service, Washington, D.C.: The Library of Congress, April 12, 1989.

Mowery, David C., and Nathan Rosenberg, *The Japanese Commercial Aircraft Industry Since 1945: Government Policy, Technical Development, and Industrial Structure*, Northeast Asia–United States Forum on International Policy, Stanford, Calif: Stanford University, April 1985.

Nihon Keizai Shimbun, January 26, 1993.

Opall, Barbara, and Naoaki Usui, "Japan, U.S. Pursue Ballistic Missile Defense," *Defense News*, October 4–10, 1993.

Prestowitz, Clyde, "Giving Japan a Handout," *Washington Post*, January 29, 1989.

Samuels, Richard, and Benjamin Whipple, "Defense Production and Industrial Development: The Case of Japanese Aircraft," in Chalmers Johnson, Laura D'Andrea Tyson, and John Zysman, eds., *Politics and Productivity: The Real Story of Why Japan Works*, Ballinger, 1989.

Sanger, David E., "U.S.-Japan Anti-Missile Plan Is on the Drawing Board," *International Herald Tribune*, September 20, 1993.

Tamama, Tetsuo, Japan Defense Research Council, Statement at the U.S./Japan Economic Agenda's *Conference on High Tech-*

nology Policy-Making in Japan and the United States: Case Studies of the HDTV and FSX Controversies, Georgetown University, Washington, D.C., June 8, 1993.

Towle, Michael D., "Jockeying for Technology," *Fort Worth Star-Telegram*, February 27, 1993.

U.S. General Accounting Office, *Japanese Aircraft Industry*, GAO/NSIAD-92-178, Washington, D.C.: U.S. Government Printing Office, June 10, 1992.

U.S. General Accounting Office, *U.S.-Japan FS-X Codevelopment Program: Statement of Frank C. Conahan, Assistant Comptroller General*, GAO/T-NSIAD-89-31, May 11, 1989.

"U.S.-Japan Anti-Missile Plan Is on the Drawing Board," *International Herald Tribune*, September 20, 1993.

U.S. House of Representatives, Committee on Foreign Affairs, *United States–Japanese Security Cooperation and the FSX Agreement: Hearings and Markup*, Washington, D.C.: U.S. Government Printing Office, 1989.

U.S. Senate, Committee on Foreign Affairs, *United States–Japanese Security Cooperation and the FSX Agreement: Hearings and Markup*, Washington, D.C.: U.S. Government Printing Office, 1989.

Wanstall, Brian, and J. R. Wilson, "Air Combat Beyond the Stall," *Interavia*, May 1990.

"X-31: The Wonder Plane," *Asian Defence Journal*, June 1990.